NUTRITION AND MALNUTRITION
Identification and Measurement

ADVANCES IN EXPERIMENTAL MEDICINE AND BIOLOGY

Recent Volumes in this Series

Volume 40
METAL IONS IN BIOLOGICAL SYSTEMS: Studies of Some Biochemical and
Environmental Problems
Edited by Sanat K. Dahr • 1973

Volume 41A
PURINE METABOLISM IN MAN: Enzymes and Metabolic Pathways
Edited by O. Sperling, A. De Vries, and J. B. Wyngaarden • 1974

Volume 41B
PURINE METABOLISM IN MAN: Biochemistry and Pharmacology of Uric Acid Metabolism
Edited by O. Sperling, A. De Vries, and J. B. Wyngaarden • 1974

Volume 42
IMMOBILIZED BIOCHEMICALS AND AFFINITY CHROMATOGRAPHY
Edited by R. B. Dunlap • 1974

Volume 43
ARTERIAL MESENCHYME AND ARTERIOSCLEROSIS
Edited by William D. Wagner and Thomas B. Clarkson • 1974

Volume 44
CONTROL OF GENE EXPRESSION
Edited by Alexander Kohn and Adam Shatkay • 1974

Volume 45
THE IMMUNOGLOBULIN A SYSTEM
Edited by Jiri Mestecky and Alexander R. Lawton • 1974

Volume 46
PARENTERAL NUTRITION IN INFANCY AND CHILDHOOD
Edited by Hans Henning Bode and Joseph B. Warshaw • 1974

Volume 47
CONTROLLED RELEASE OF BIOLOGICALLY ACTIVE AGENTS
Edited by A. C. Tanquary and R. E. Lacey • 1974

Volume 48
PROTEIN–METAL INTERACTIONS
Edited by Mendel Friedman • 1974

Volume 49
NUTRITION AND MALNUTRITION: Identification and Measurement
Edited by Alexander F. Roche and Frank Falkner • 1974

NUTRITION AND MALNUTRITION

Identification and Measurement

Edited by

Alexander F. Roche and Frank Falkner

The Fels Research Institute
Yellow Springs, Ohio

PLENUM PRESS • NEW YORK AND LONDON

Library of Congress Cataloging in Publication Data

Burg Wartenstein Conference on Physical Anthropology and Nutritional
Status, 1973.
Nutrition and malnutrition.

(Advances in experimental medicine and biology, v. 49)
Includes bibliographies.
1. Nutrition—Congresses. 2. Malnutrition—Congresses. 3. Anthropometry—
Congresses. I. Roche, Alexander F., 1921-   ed. II. Falkner, Frank Tardrew,
1918-   ed. III. Title. IV. Series. [DNLM: 1. Anthropology, Physical—
Congresses. 2. Nutrition—Congresses. 3. Nutrition disorders—Congresses.
W1 AD559 v. 49 1973 / QU145 B954n 1973]
TX345.B8 1973   641.1   74-13950
ISBN 0-306-39049-3

Proceedings of the Burg Wartenstein Conference on Physical Anthropology
and Nutritional Status held August 6-15, 1973

© 1974 Plenum Press, New York
A Division of Plenum Publishing Corporation
227 West 17th Street, New York, N.Y. 10011

United Kingdom edition published by Plenum Press, London
A Division of Plenum Publishing Company, Ltd.
4a Lower John Street, London W1R 3PD, England

Printed in the United States of America

Preface

The Burg Wartenstein Symposia have become rightly celebrated for achieving their original purpose: to be of service and interest to the anthropological profession as a whole and to contribute to related sciences. We are specially grateful to the Board of Directors of the Wenner-Gren Foundation for Anthropological Research for the award of a symposium on Physical Anthropology and Nutritional Status. We had considered that such a subject was timely and that an inter-disciplinary approach would contribute useful knowledge in a most important area--particularly in the field of child health. This publication of the proceedings will show the degree of success of these aims.

Mrs. Lita Osmundsen, Director of Research at the Wenner-Gren Foundation, not only steered us in the early stages but, during our delightful time at Burg Wartenstein, and subsequently, she has been that most charming of crosses--den mother and first class science administrator. We are deeply grateful to her. And grateful, too, for the organization and friendly warm spoiling by the Wenner-Gren staff in residence at Burg Wartenstein. Our nutritional status was high. Nothing was too much trouble for them and it was a wrench to leave and say goodbye to them.

The Foundation has since supported our efforts towards this publication. Here we owe a very special debt of gratitide to the cooperation and friendly expertise of Mr. Seymour Weingarten, Senior Editor of the Plenum Publishing Corporation, aided by Mr. John Matzka, Managing Editor of this Corporation.

v

Finally, some of us were originally doubtful about the prognosis of being sequestered in an Austrian castle--however beautiful--for one week with some 20 scientific colleagues. Those having these thoughts could not have been proved more wrong. We are grateful, above all, to the participants for their contributions, then and since, and indeed for making the outcome so happy, healthy and lusty.

Alex F. Roche

Frank Falkner

Co-Chairmen

Contents

Jelliffe, E.F.P. and Gurney, M.: Definition Of
 The Problem   1

Yarbrough, C., Habicht, J.-P., Martorell, R. and
 Klein, R.E.: Anthropometry As An Index Of
 Nutritional Status   15

Terada, H.: A New Apparatus For Stereometry:
 Moiré Contourograph   27

Cheek, D.B., Brayton, J.B. and Scott, R.E.:
 Overnutrition, Overgrowth And Hormones
 (With Special Reference To The Infant Born
 Of The Diabetic Mother)   47

Metcoff, J.: Maternal Leukocyte Metabolism In
 Fetal Malnutrition   73

Paŕísková, J.: Interrelationships Between Body
 Size, Body Composition And Function   119

Brǒzek, J.: From QUAC Stick To A Compositional
 Assessment Of Man's Nutritional Status . . .   151

Widdowson, E.M.: Changes In Pigs Due To
 Undernutrition Before Birth, And For One,
 Two, And Three Years Afterwards, And The
 Effects Of Rehabilitation   165

Chow, B.F.: Effect Of Maternal Dietary Protein
 On Anthropometric And Behavioral Development
 Of The Offspring   183

Goldstein, H.: Some Statistical Considerations
  On The Use Of Anthropometry To Assess
  Nutritional Status 221

Wolański, N.L.: Biological Reference Systems
  In The Assessment Of Nutritional Status 231

Buzina, R. and Uemura, K.: Selection Of The
  Minimum Anthropometric Characteristics To
  Assess Nutritional Status 271

Johnston, F.E.: Cross-Sectional Versus
  Longitudinal Studies. 287

Jelliffe, D.B.: Relative Values Of Longitudinal,
  Cross-Sectional And Mixed Data Collection In
  Community Studies 309

Cravioto, J. and DeLicardie, E.R.: The Relation
  Between Size At Birth And Preschool Clinical
  Severe Malnutrition 321

Malcolm, L.A.: Ecological Factors Relating To
  Child Growth And Nutritional Status 329

Conference Participants. 353

Index. 355

DEFINITION OF THE PROBLEM

E. F. Patrice Jelliffe and Michael Gurney

From the University of California, Los Angeles,
California, USA, and Caribbean Food and Nutrition
Institute, Kingston, Jamaica

The gross dimensions of the problem facing workers
in the nutritional field are well known, but definition
of the problem is difficult.

"Nutritional status" per se is an abstract concept
and to assess it, workers have had at their disposal
methods, measurements and techniques that are still being
criticized, partially accepted, refined and continuously
evaluated. Such relatively slow progress would be considered inefficient in any non-medical field. However,
the ever changing aspects of the complex problem of human
malnutrition, with its multiple interrelationships, cannot
be underrated by experienced nutrition workers.

To diagnose accurately the problems of community
nutrition, a need exists for suitable methods and measures that can assess accurately, simply and inexpensively,
the nutritional status of populations at risk. Recent
concern with problems of community nutrition has clearly
indicated this need to be able to define the extent of
malnutrition in affected communities. Newer methods and
techniques have been tried, particularly in the past
decade, to measure the results of dietary deficiencies
(and excesses) of total calories (energy) and of protein,
especially in young children. These methods have been
concerned with the assessment of the prevalence and

1

incidence of protein-calorie malnutrition in early childhood, but also with the apparently increasing problem of
infantile obesity. The assessment of the prevalence of
malnutrition is an essential part of community nutrition
surveillance.

The measurement of the extent of the problem is
needed for various reasons. First, such assessments of
prevalence can be used as baselines to measure the effectiveness of planned intervention programs, or of un-
planned changes in the particular community, or of secular trends.

Data showing the prevalence of malnutrition and its
geographical, economic and social distribution within a
country or region are plainly of the greatest importance
in suggesting how limited resources can be deployed to
areas of greatest need. In addition, the availability
of hard figures, presented in an understandable form,
showing the commonness of malnutrition in a particular
community can be valuable in trying to stimulate the
provision of the necessary action and resources from
fund-controlling administrators and politicians.

The need for methods of assessing malnutrition,
particularly in young children, has come very much into
focus as a result of other large-scale studies on a
variety of important topics, including the relationship
between mental development, immunity and nutritional
status.

Lastly, on an individual basis, there has been an
increasing awareness of the need to monitor growth to
detect children with early malnutrition and to screen
those at risk as early as possible. This has been undertaken by using various forms of growth charts, such as
the one introduced and popularized by Morley.[1] The latter
type of chart not only assesses the growth progress of
the child under surveillance but is also an excellent
teaching aid for medical and paramedical personnel, as
well as the child's parents, particularly the mother.

METHODS OF NUTRITIONAL ASSESSMENT

The nutritional assessment of human communities can be undertaken in various ways. These have been divided into three main groups of approach--<u>direct</u>, <u>indirect</u>, and <u>ecological</u>.[2] The present concern will be mostly with the direct assessment of nutritional status in human communities. This can be considered under the four headings--<u>clinical signs</u>, <u>biochemical tests</u>, <u>tissue tests</u> and <u>anthropometry</u>. Each of these approaches has its limitations as a diagnostic tool. The most appropriate blend of these methods will have to be used in surveys concerned with the etiology and treatment of malnutrition and the evaluation of the success of intervention programs.

Clinical Signs

It is increasingly appreciated that clinical signs, which should be recorded selectively to give a general over-view of the local situation, are very variable and highly subjective[3] and depend greatly on the experience of the clinician. They need to be supplemented by other data.

Biochemical Tests

These include amino acid inbalance, serum albumin, blood vitamin levels, hydroxyproline excretion, and urinary creatinine. They require samples of blood or urine to be taken in the field. These samples must be preserved and transported for analysis in a specialized laboratory. Finally, the results must be interpreted correctly.[2] None of these steps is easy, and recently there has been an increasing tendency to use such tests selectively, and sometimes only on a subsample, because of their cost and the difficulties encountered under field circumstances.

Tissue Tests

Several of these have been introduced in recent
years, such as examination of hair root morphology.[4]
Samples of hair are easy to collect from children,
easily stored, transported and examined. Hair, as a
sensitive protein tissue, indicates the early effects
of protein malnutrition by changes in bulb root diameter
when serum albumin levels in the blood are still within
normal limits. The results of this test can be interpreted without knowledge of the precise age. The findings are useful if combined with anthropometric measures, [5,6] such as arm circumference, but used alone
they would not yield sufficient information.

Anthropometry

This discipline provides measures of body size and
shape, and also indicates the dimensions of some body
compartments. By general consensus of those working in
this field, nutritional anthropometry has a most significant role in the direct assessment of nutritional status
in communities, especially in young children, where the
problem of malnutrition is most severe and extensive.
In addition, anthropometry may play a considerable role
in the nutritional assessment of older children and
adults.

It must be appreciated, however, that the growth of
children is influenced by a variety of factors such as
socioeconomic status of the parents and the sex and birth-
rank of the child. Climate, seasonal variations, infections, parasites and psychological factors also, directly
or indirectly, affect the child's nutritional status.

PROBLEMS OF ANTHROPOMETRIC ASSESSMENT

The anthropometric assessment of nutritional status
is concerned principally with the detection of protein-
calorie malnutrition of early childhood (PCM), especially
in cross-sectional surveys in less developed countries.
However, as with the other methods, there are problems

in the selection of both measurements and methods and
their standardization, in the assessment of age and in
the selection of reference standards.

Selection of Measurements and Methods

The selection of measurements and methods that must
be simple and replicable, using suitably trained staff
and accurate apparatus, will depend on the information
sought and the funds and staff available. In addition,
when the problem of PCM is considered, accurate knowledge
of the child's age is a key factor in the selection of
measurements.

Usually PCM is the main problem to be assessed by
anthropometric means. Four measurements are essential--
weight, height (length), arm circumference and triceps
fatfold. In addition, head and chest circumferences are
frequently included in the "basic six." The measurements
included will depend also on the objectives of the survey and considerations including logistics and the distance from the home-base. For example, in a famine or
disaster area children can be screened by measurement of
the arm circumference alone,[7] or by the QUAC stick which
groups children on the basis of arm circumference for
height.

Recent studies suggest that different measurements
may be interpreted in relation to not only the severity
of PCM, but also to its acuteness or chronicity, and to
its type, in relation to depletion of calorie stores
(fat) or of protein stores (muscles).[8] A low body weight
or arm circumference is associated with a relatively
acute or recent onset of malnutrition. By contrast, a
low height-for-age indicates malnutrition of some chronicity. The arm circumference can be analyzed, together
with the triceps fatfold to permit separate analysis of
the fat and muscle cross-sectional areas, indicating
whether the deficiency is principally of protein or
calories.[9]

While differentiation of PCM into various categories
seems possible with the measurements previously mentioned,

there is a need for confirmation of their usefulness in
diverse ecologies. In addition, an unsolved problem concerns the possible duplication of information obtained
by these different measurements. In the limited time
available in surveys, unnecessary duplication must be
eliminated.

Standardization of methods and techniques is necessary to ensure comparability of results among countries,
and to make comparisons possible between surveys made at
different times within the same community. Currently
such comparisons are extremely difficult and unreliable
because differences exist among the techniques used in
surveys in various parts of the world and because there
is a lack of agreement as to which of the available reference standards should be used for comparison.

There is a need to standardize anthropometric techniques but debate continues concerning methods and prac-
tices. Standard techniques should be followed and the
precise details of the methods used, including the apparatus, should be documented and reported in all studies.

Usually only limited funds are available to undertake
community studies to investigate nutritional status.
Therefore, the most economical and simplest measurements
and equipment, that can give scientifically valuable and
replicable results, must be identified. Consideration
has to be given to the cost of the equipment and staff
needed to prepare and manage the survey, and to analyze
and interpret the data. Anthropometry can be undertaken
in developing countries employing specially trained
supervised para-medical personnel. The time taken to
undertake various types of measurements needs consideration. For example, the QUAC stick method can be applied
in only a fraction of the time needed to weigh and measure
the length of a preschool child.

Likewise, equipment must be sturdy, simple, inexpensive and easily transportable. Many different types are
used in various parts of the world making the standardization of methods a difficult task. However, a recent
survey into types of commonly used and available scales
showed that for mobile field surveys of young children,

a portable Salter spring balance gave sufficiently
accurate results and was relatively cheap and easily
transportable. Considerations of costs are important in
the selection of anthropometric equipment, including
length boards and fatfold calipers. One advantage of
the measurement of arm circumference is that the equipment required is only a non-stretch centimeter tape
measure.

Availability of Precise Chronological Age

If the precise age of young children is known, the
selection of measurements and their interpretation become
much easier. In most parts of the world, the precise
ages of young children are unknown, because they are not
of social significance. Also, in many communities, docu-
mentation regarding the correct age is not available.
Use can be made of horoscopes, birth certificates or
baptism certificates, if such documents exist.

When the precise age of the young child cannot be
ascertained, it may only be possible either to estimate
the age to the nearest year, or to classify children
roughly as preschool children (one to five years). This
complicates the choice of measurements to be made and
makes their interpretation more difficult.

Under these circumstances, it is essential to use
an age-independent ratio, which gives information concerning nutritional status (i.e., chest circumference/head
circumference), or to compare a measurement which reflects
a nutritionally labile body tissue (body weight or arm
circumference),[7] with a measurement of body tissue that
is much less affected by malnutrition, and which cannot
decrease (height, arm length or head circumference).[10]
Attempts have been made to define a "dental second year"
as an approximate measure of age to test the validity of
a low chest circumference/head circumference ratio, or
arm circumference as field tests for PCM.[11]

Many composite indices have been used in field surveys during the past two decades; following the pioneering
example of some psychologists who, in the 1950's wished

to extrapolate meaningful results from multiple tests
applied to their patients and used statistical techniques
of multivariate analysis. Such techniques aim to replace,
numerous correlated measurements by a smaller number of
uncorrelated linear combinations of these measurements.

Among the composite indices were, to name but a few,
the scoring system, [12,13] the point system[14] (modified
by Timmer[15]), and Kanawati's "index of thriving"[16] which
uses the sum of scores of four anthropometric measures,
(weight, midarm circumference, height or length, and head
circumference). Furthermore, Burgess et al.[17] and
Moffat[18] assessed PCM on the basis of three or more signs
including one abnormal anthropometric measurement. When
composite anthropometric measurements are used, the component measures must be simple to collect, and must each
measure an unrelated major component of the growth failure,
affecting different body tissues and physical dimensions,
that is seen in the various forms of protein calorie malnutrition.

Hanafy et al.[19] used the <u>nutritional state index
of the infant</u> (N.S.I.I.). This is based on weight and
arm circumference expressed as percentages of predicted
values and serum albumin as a percentage of normal (3.8
g/100 ml) and the nutritional state index of the mother
(N.S.I.M.). The latter is based on weight, arm circumference, serum albumin, urea nitrogen and creatinine
nitrogen.

In recent years, however, emphasis has been given
to anthropometric measurements that are truly age independent, provided the children can be grouped as being
of preschool age (one to five years). These measurements
include arm circumference x 100/height,[20-22] weight/
height, which Gopalan[22] found was highly correlated with
undernutrition in young children as judged by a low weight
for age or by the presence of clinical signs. Other
indices include weight/height[1.6], arm circumference/
stature[23] and arm circumference/length of upper arm x
100.[24] The latter index possibly represents environment
(nutrition)/genetic variables. Arm circumference alone
has been employed in surveys, using an approximate standard for the whole age range from one to five years. Using

the concept of indices, problems occur in relation to the interpretation of differences and their comparability. Much work remains to be done in this area.

Standards of Comparison

Physical growth, both of the body frame and total body weight, depends upon many interacting influences which include genetic and environmental ones, particularly nutrition. The major and, as yet, unsolved difficulty is to differentiate precisely between genetic and environmental influences: between "heritability" and "ecosensitivity."[25]

For example, recent work in various parts of the world has indicated that young children of well-to-do parents, who have been well fed and protected from unnecessary infections, frequently have similar dimensions to those derived from Caucasian children in the USA or Europe. Eksmyr found that the anthropometric measurements of preschool children of the Ethiopian élite in Addis Ababa were similar to those for North American children.[26] Similar findings have been reported from Kuala Lumpur in Malaya.[27]

Conversely, it is apparent that genetic differences in body physique occur, and some of these are extreme, for example, between the pygmies and the Watutsi, or between the big chested Andean populations and linear Nilotics.[28] Less obvious, is the finding that "normal" triceps fatfold seems to be thinner in at least some African populations, in comparison with other body fatfolds. This difference may be genetically determined.

A further problem with regard to standards is that those currently used, for example, those in the appendices to WHO Monograph #53[2] were derived from different populations, at different times, by different observers, using different techniques and with different types of apparatus. At best, they have the advantage of being internationally available, so that they can be used for comparisons between one community and another. From the point of view of nutritional anthropometry in young children, it is

particularly unfortunate that the standards available
for the commonly used measurements--weight, height
(length), arm circumference, triceps fatfold, circumferences of head and chest--are only useful approximations.

Another difficulty with regard to reference standards
is that, until recently, it was widely considered that
the larger the better, or, in other words, the greater
the growth the healthier the child. The fallacy of this
view has been increasingly endorsed by the recent recognition in the United States and Europe that obesity in
early childhood is increasing and appears to be related
primarily to over-feeding, in other words over-nutrition.[29]
This poses considerable problems because it is difficult
to define the commencement of over-nutrition either intrauterine (as described by Chow--personal communication--in
animal experiments) or in the early days of life. On a
practical basis, it is difficult to judge whether particular reference standards, developed in recent decades, were
derived from children before or during the modern vogue
for "double feeding" young children. Frequently over-
concentrated cow's milk formulas are given in quantities
suggested by the mother's ideas rather than the baby's
needs, and processed solid foods are introduced in the
early weeks of life. Difficulties in this regard are
indicated by the results from a recent study of infants
attending a well-baby clinic in Northern England, when
17 percent were found to be obese (120%-140% of standards).[30]

Ideally, there is a need to define not maximal growth
standards, but rather optimal or desirable standards,
which would, of necessity, be difficult and time consuming
to collect because they would have to take into account
not only growth in early childhood, but also intellectual
development and resistance to infection, as well as the
long-term sequelae, including the incidence of cardiac
disease and diabetes, and longevity.[31] In other words,
anthropometric variables may reflect factors that influence growth and, additionally, they may be associated with
health related effects of unusual growth. This type of
undertaking, which would be gigantic in scope and tre-
mendously and literally time consuming, does not seem

currently practical or likely to be undertaken.

From a practical point of view, there is a real need, as has been outlined by the Falkner Committee, to undertake the collection of modern standards based on measurements made by agreed and standardized techniques in different representative areas in the world.[22] While it would not be possible to undertake such a study in each country in all genetic pools, the aim should be to reach prior agreement as to the definition of the characteristics of the so-called élite group and, at the same time, to ensure that each survey in such a multi-national study record and report a precise detailed description of those being measured.

Plainly, a basic problem lies in defining the optimal population to be measured in each main genetic pool. In the past, so-called "standards" have been correctly criticized when derived from, for example, data recorded in a child welfare clinic in a developing country. Many of these children would be far from optimal because such clinics see a high percentage of children suffering from a continuous procession of infections and parasitic diseases. Conversely, in relation to the élite group in developing countries, a major and as yet unsolved problem is how to ensure that the measurements are not being made on overnourished children. This would have the double ill effect of setting undesirably high reference standards for the particular community, and, at the same time, basing the standards on measurements that could be pathologically excessive.

A basic need is, therefore, to define more precisely how the "optimal" would be sought in different genetic pools in representative parts of the world. This approach is one of the greatest practical urgency, because, as indicated earlier, anthropometry represents the most hopeful method of assessing different degrees and forms of protein-calorie malnutrition and obesity in early childhood.

The need for up-to-date, relevant standards, based on new data from representative ecologies, is plainly a world priority in practical community nutrition.

REFERENCES

1. Morley, D.C.: A health and weight chart for use in
    developing countries. Trop. Geog. Med., 20:101,
    1968.

2. Jelliffe, D.B.: The assessment of the nutritional
    status of the community (with special reference
    to field surveys in developing regions of the
    world). WHO Monogr. Ser., 53:1, 1966.

3. International Union of Nutritional Sciences:
    Assessment of protein nutritional status. A
    Committee Report. Amer. J. Clin. Nutr., 23:807,
    1970.

4. Bradfield, R.B.: A rapid tissue technique for the
    field assessment of protein-calorie malnutrition.
    Amer. J. Clin. Nutr., 25:720, 1972.

5. Bradfield, R.B. and Jelliffe, E.F.P.: Early assess-
    ment of malnutrition. Nature, 225:283, 1970.

6. Bradfield, R.B., Jelliffe, E.F.P. and Neill, J.:
    A comparison of hair root morphology and arm
    circumference as field tests of protein-calorie
    malnutrition. J. Trop. Pediat., 16:195, 1970.

7. Jelliffe, E.F.P. and Jelliffe, D.B. (Eds): The arm
    circumference as a public health index of protein-
    calorie malnutrition of early childhood. J. Trop.
    Pediat., 15:177, 1969.

8. Gurney, M., Jelliffe, D.B. and Neill, J.: Anthro-
    pometry in the differential diagnosis of protein-
    calorie malnutrition. J. Trop. Pediat., 18:1,
    1972.

9. Gurney, J.M. and Jelliffe, D.B.: Arm anthropometry
    in nutritional assessment: Nomogram for rapid
    calculation of muscle circumference and cross-
    sectional muscle and fat areas. Amer. J. Clin.
    Nutr., 26:912, 1973.

10. Jelliffe, D.B. and Jelliffe, E.F.P.: Age independent anthropometry. Amer. J. Clin. Nutr., 24:1377, 1971.

11. Jelliffe, D.B.: Field anthropometry independent of precise age. J. Pediat., 75:334, 1969.

12. Jelliffe, D.B. and Welbourn, H.F.: Clinical signs of mild-moderate PCM. Swedish Nutrition Foundation, Symposium No. 1, Uppsala, Almquist and Wiksells, 1963.

13. McLaren, D.S., Pellet, P.L. and Read, W.W.C.: A simple scoring system for classifying the severe forms of protein-calorie malnutrition of early childhood. Lancet, i:533, 1967.

14. Oomen, H.A.P.C.: The external pattern of malnutrition in Djakarta toddlers. Doc. Med. Geog. Trop., 7:1, 1955.

15. Timmer, J.M.: Child malnutrition in Jogjakarta, Java. Degree and single symptoms in age groups. Trop. Geog. Med., 17:126, 1965.

16. Kanawati, A.A. and McLaren, D.S.: Assessing the growth and nutrition of young children. Trans. Roy. Soc. Trop. Med. Hyg., 62:569, 1968.

17. Burgess, H.J.L., Maletnlema, N.T. and Burgess, A.P.: Nutrition survey in Tabora, Tanzania. Trop. Geog. Med., 21:39, 1969.

18. Moffat, M.: Mobile young child clinics in rural Uganda: A report on the Ankole preschool protection programme (1967-69). Mimeographed.

19. Hanafy, M.A., Morsey, N.R.A., Siddick, Y., Habib, Y.A. and Lozy, M.: Maternal nutrition and lactation performance. J. Trop. Pediat., 18:187, 1972.

20. Klerks, J.V.: The AHC index. Berita Kementerian Kesehatan Indonesia. 5:11, 1956.

21. Rao, K.V. and Singh, D.: An evaluation of the relationship between nutritional status and anthropometric measurements. Amer. J. Clin. Nutr., 23:83, 1970.

22. International Union of Nutritional Sciences: The creation of growth standards. A Committee Report. Amer. J. Clin. Nutr., 25:218, 1972.

23. Arnhold, R.: (XVII) The QUAC stick: A field measure used by the Quaker Field Service Team, Nigeria. J. Trop. Pediat., 15:243, 1969.

24. Wolański, N.: Personal communication, 1969.

25. Hiernaux, J.: Heredity and environment: Their influence on human morphology. A comparison of two independent lines of study. Amer. J. Phys. Anthrop., 21:575, 1963.

26. Eksmyr, R.: Anthropometry in privileged Ethiopian preschool children. Acta Paediat. Scand., 59:157, 1970.

27. McKay, D.A., Lim, R.K.H., Notaney, K.H., and Dugdale, N.E.: Nutritional assessment by comparative growth achievement in Malay children below school age. Bull. World Hlth. Org., 45:233, 1971.

28. Frisancho, A.R.: Human growth and pulmonary function of a high altitude Peruvian Quechua population. Human Biol., 41:365, 1969.

29. Jelliffe, D.B. and Jelliffe, E.F.P.: Fat babies-- prevalence, perils and prevention. In press, 1974.

30. Shukla, A., Forsyth, H.A., Anderson, C.M. and Marwah, S.M.: Infantile overnutrition in the first year of life. Brit. Med. J., ii:507, 1972.

31. Walker, A.R.P. and Richardson, B.D.: International and local growth standards. Amer. J. Clin. Nutr., 26:897, 1973.

ANTHROPOMETRY AS AN INDEX OF NUTRITIONAL STATUS*

Charles Yarbrough, Jean-Pierre Habicht,
Reynaldo Martorell and Robert E. Klein

From the Division of Human Development,
Institute of Nutrition of Central America and
Panama (INCAP), Guatemala City, Guatemala,
Central America

Our present concern is to address the problem of
the use of anthropometry as an index of nutritional
status. Inferences drawn from such usage must be based
upon knowledge of the accuracy and validity with which
anthropometric indices measure nutritional status. To
illustrate this problem, we will refer to one particular
anthropometric index of nutritional status, weight, the
most widely used index, and only to preschool children.
We are concerned particularly with those aged between
two and three years who are at most risk of protein-
calorie malnutrition in the villages where we work,[1,2]
because they have just been weaned.

Consideration will not be given to the relationships
between anthropometric variables and functional outcomes,
e.g., cellular function, mental development, illness
although there is a need for the investigation of these
relationships. These investigations should seek to determine not only which anthropometric variables are good
indicators of functional outcomes but the extent to which

* This research was supported by Contract #PH 43-65-640
  from the National Institute of Child Health and Human
  Development, National Institutes of Health, Bethesda,
  Maryland, USA

functional outcomes are determined by nutritional status.
Consequently, non-anthropometric variables concerning,
for example, nutrient intake or cellular metabolism should
be studied also.

In summary, we will present evidence that attained
body weight is a good index of nutritional status for
protein-calorie malnutrition at the extremes, i.e.,
clinical or severe protein-calorie malnutrition and over-
fed children. But through the middle range of mild to
moderate protein-calorie malnutrition, attained weight
is not sufficiently sensitive to changes in nutritional
status to be used <u>alone</u> as an index of protein-calorie
malnutrition. Neither are changes in weight in the same
child sensitive to changes in nutrition over the range
of mild to moderate malnutrition. Thus body weight, used
alone, does not appear to be a useful field measurement
to study the determinants of protein-calorie malnutrition
in rural underdeveloped areas. However, it is valuable
for screening to identify children with severe protein-
calorie malnutrition who need medical and nutritional
intervention. Body weight is useful also as a population
index of protein-calorie malnutrition--its major limitation is "reliability," which large sample size copes with
nicely.

We came to these conclusions by attempting to answer
the following questions:

1. How specific are changes in body weight for
    changes in protein-calorie malnutrition?

2. How sensitive is body weight to changes in
    protein-calorie malnutrition?

Various levels of specificity should be considered
in relation to an index of nutritional status. In the
case of body weight, the first level of specificity is
to determine how well we can measure body weight. The
next level is to ask how well body weight reflects an
underlying structure more directly related to nutrition,
namely, body composition? Finally, with what specificity
are changes in body composition related to changes in
nutritional status?

The quantitative relationships between nutritional status, body composition and weight are not essential to our understanding of body weight as an index of malnutrition but the conceptualization of these relationships is important. For instance, changes in body weight can reflect changes in hydration, or the contents of the bladder or gut which are not related to changes in body protein mass or caloric stores. A measure of the importance of this variability, which does not reflect protein-calorie nutrition, can be estimated from short term variations in body weight (Table I). This day-to-day variation in body weight disturbs the estimate of underlying protein mass and caloric stores. In the context of mild to moderate malnutrition, these short term changes in body weight are non-nutritional in character. It is important to be explicit about this last statement because a pediatrician concerned with a child's water balance would

TABLE I

Precision of Measurement and Variability of Body Weight

Component of Variability	Measured by	Result	Value of Component	Percent of Variance
Precision of instrument	Repeated measurement of metal standards	$\pm$ 3 g	$\pm$ 3.0 g	0.02
Precision of child-measurements	Repeated measurement of child within 2 hrs	$\pm$20 g	$\pm$19.8 g	0.97
Short-term variations within child	Repeated measurement from day to day	$\pm$201 g	$\pm$200.0 g	99.01
		Total	$(201.0 \text{ g})^2 =$	100.00

consider hourly variations in body weight as extremely
informative of the child's water nutritional status.
Similarly, in a study of acute total starvation day-to-
day variations in body weight would be informative.

Relative to the day-to-day variations in body weight,
the specificity of the measurement of body weight itself
is excellent. The contributions of instrument and measuring technique variability to the total non-nutritional
variance is negligible (less than one percent). The day-
to-day variability in two to three year olds due to hydration and bladder and gut contents is, in fact, so large
that an instrument precision of 100 g instead of 10 g
would only increase the total non-nutritional standard
deviation of body weight by 2 g to $\pm$ 202 g. This is
important because it means that one may confidently sacrifice instrument precision for increased robustness of
scales in field studies of protein-calorie malnutrition.

Despite a high non-nutritional variability, body
weight promises to be useful if the variability across
children is nutritional in character. Thus, Table II presents the expected sample sizes needed to show significant associations between attained body weight and another
variable without non-nutritional variability, if most of
the variability across children is due to differences in
nutrition. Similarly, if the differences in body weight
are due mostly to differences in nutrition, a difference
of 1.34 kg will separate the means of the upper and lower
halves of the population. One would only need 40 children
in such a population for statistical significance (P_a < .20
and P_b < .05), that is, 80 percent of the times when no
statistically significant association is found at the
five percent level there is indeed no such association
and 95 percent of the times a statistically significant
association is found this will reflect a true association.
Table II contains estimates of the number of children in
each of two groups needed to demonstrate a statistical
difference in weight of 1 kg at three years of age. The
value of 1 kg was selected because it is about half the
difference between the mean weight of the Ladino rural
Guatemalan children in the INCAP study villages[3] and the
mean weight of Denver (USA) children.[4]

Two questions now arise:

1. What evidence do we have that most of the $\pm$ 1263 g difference in the population is nutritional?

2. What evidence do we have that this variability in weight is not solely due to variability in height?

The second question is easy to answer. Table III shows that about $\pm$ 700 g is not due to differences in height, which means that about $\pm$ 1050 g was due to differences in height. The expected sensitivity of weight in children of similar stature is, however, still good if the $\pm$ 700 g represents nutritional differences among

TABLE II

Sensitivity of Body Weight as a
Measure of Nutritional Status

Attained Body Weight:	
Day to day variability	$\pm$ 201 g
Sex-specific variability at 3 years	$\pm$1263 g
Maximum possible correlation with another variable and	$r=.987$
Number of children for $P_a<.20$; $P_b<.05$	4
Number of children in each of 2 groups to show differences in attained weight at 3 years of 1,000 g for $P_a<.20$; $P_b<.05$	20

P_a and P_b refer to the powers of statistical tests of significance (see text).

children.   What evidence do we have on that score?

There are various types of evidence. First, if we consider the weight for height graph of the children in the study villages, it can be seen that although these children are stunted in weight by about 15 percent[3] they have almost the same weight for height (Figure 1) as the United States children included in the Denver sample. Secondly, these children showed increased growth in stature after protein supplementation,[5] but there was no change in weight for stature among these supplemented children.   These findings from the INCAP study villages in Guatemala indicate that little of the $\pm$ 700 g difference in weight is nutritional in origin in these mild to moderately malnourished children.

TABLE   III

Sensitivity of Body Weight Relative to Stature
as a Measure of Nutritional Status

Attained body weight:	
Day to day variability	$\pm$201 g
Height-specific variability at 2-3 years	$\pm$700 g
Maximum possible correlation with another variable	r=.960
and Number of children for $P_a<.20$; $P_b<.05$	5
Number of children in each of two groups to show a difference in attained weight for stature of 1,000 g for $P_a<.20$; $P_b<.05$	7

P_a and P_b refer to the powers of statistical tests of significance (see text).

Another way of studying the role that nutrition can play in the determination of weight is to study weight changes across a broad range of grades of protein-calorie malnutrition. Table IV shows the approximate differences between four groups of children with different protein-calorie nutritional status: a group hospitalized with kwashiorkor, who had just lost their visible edema ("clinical PCM"); a group recuperating from protein-calorie malnutrition ("moderate PCM"); a group similar in height and weight to those of the study villages and who were hospitalized for operations or short term infections ("mild PCM"); and finally a group of well-fed children. The differences in weight of children of similar stature between the well-fed and the clinically malnourished group or between the middle two groups and the two extremes are so striking that they hardly require statistical testing. On the other hand, the difference between the children still hospitalized for malnutrition and those who were hospitalized for other reasons is very small, and would be difficult to demonstrate. Thus, body weight is a good indicator of imminent clinical malnutrition but

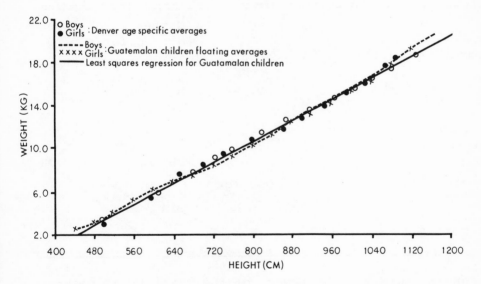

Figure 1. Weight-height relations in Guatemalan and Denver children from birth through six years.

is a poor indicator of protein-calorie malnutrition, within the ranges usually found in field studies.

So far we have discussed only the use of attained body weight as an index of nutritional status. Also, we have investigated the use of <u>increases</u> in body weight as an index of nutritional status. Table V indicates that the sensitivity of body weight increments is a little lower than that of attained body weight. If, by supplementation, we can improve the nutritional status of the village children so that they grow at the same rate as United States children, how long would we have to wait to see a difference in growth between supplemented and unsupplemented children?

TABLE IV

Severity of Malnutrition, Weight Deficit,
and Sample Size
(Calculations for children of same height)

Degree of malnutrition	Weight deficit	Number of children needed for $P_a > .20$; $P_b > .05$
Clinical PCM	2,250 kg	
		16
Moderate PCM	1,250 kg	7
		300
		3
Mild PCM	1,000 kg	13
		13
Well-fed children	0 kg	

P_a and P_b refer to the powers of the statistical tests of significance (see text).

Table VI presents various intervals and the approximate sample sizes needed in each of the two groups, supplemented and unsupplemented, to show a statistical difference between the two groups. The results indicate clearly that weight gain in mild to moderate protein-calorie malnutrition cannot be used as a short term index of nutritional status. Only intervals of six months or longer yield growth improvements that can be measured by practicable sample sizes.

Given such a long interval, might height increments be a better index of nutritional status than is weight gain? This is, in fact, the case. To demonstrate that a 20 percent increment in growth in stature associated with supplementation is real requires only 50 children in each group if the data refer to a six-month interval.

These predictions of relative sensitivity between weight gain and gain in stature as indices of nutritional status are confirmed in practice. Almost always,

TABLE   V

Sensitivity of Body Weight Increments
as a Measure of Nutritional Status

Increment in body weight:	
Standard deviation of measurement	$\pm$284 g
Population standard deviation of the increment between 2.5 and 3 years	$\pm$640 g
Maximum possible correlation with another variable	r=.896
Number of children for $P_a<.20; P_b<.05$	6

P_a and P_b refer to the powers of statistical tests of significance (see text).

differences in the growth of preschool children, due to
differences in supplementation or differences in illness
experience, are recognized more quickly in the study villages by examining differences in gains of stature than
by using gains in height.

Similar investigations to those described above indicate that head circumference, like body weight, conveys
no information other than that provided by stature, until
some time during the second year of life. Thereafter,
there is a deficit of head circumference related to stature in children who were malnourished into the early preschool years. Head circumference may be a good index of
early protein-calorie malnutrition in the older preschool
child, but in the infant it is no more informative than
stature.

TABLE VI

Intervals of Growth in Body Weight
and Sample Size
(24-48 months)

	Standard deviation of increment measurement	± 240 g	
	Daily increase in population variance	$(\pm 42.5$ g$)^2$	
Interval	20% difference in weight gain (g)	Standard deviation of increment (g)	Approximate sample size for $P_a < .20; P_b < .05$
1 week	9	± 300	$>5,000$
1 month	34	± 370	1,240
3 months	104	± 500	310
6 months	207	± 640	130
1 year	410	± 860	67
2 years	800	$\pm 1,150$	26

Concern for sensitivity and specificity require that age be excluded as a confounding variable. A value of an indicator of nutritional status can have different interpretations with changes in age, because the mean value may alter with age or because deviations from the average may have different meanings at different ages. The meaning of an index of nutritional status depends upon the aims of the study; these will determine the reference against which an indicator of nutritional status should be tested for sensitivity and specificity. For instance, if the goal is to infer nutrient intake, then nutrient intake should be the referent. Similarly, depending upon the objective for which the indicator of nutritional status will be used, body composition, functional tests, clinical state or even death may be used as the referent. The value of an indicator of nutritional status is seldom useful without knowledge or inferences about its longitudinal history. Often this is inferred from ecological considerations in field studies, but it is sometimes forgotten when interpreting cross-sectional data.

In conclusion, we consider that, despite widespread usage, little is known quantitatively about the relationships between anthropometric variables and protein-calorie malnutrition except at the clinically apparent extremes. In this field, there is ample scope for imaginative work by clinicians, epidemiologists, statisticians and the body composition physiologists.

REFERENCES

1. Klein, R.E., Habicht, J-P. and Yarbrough, C.: Some methodological problems in field studies of nutrition and intelligence. In: D.J. Kallen (Ed) Proceedings of the Conference on the Assessment of Tests of Behavior from Studies of Nutrition in the Western Hemisphere. Washington, D.C., Government Printing Office, in press.

2. Mejía Pivaral, V.: Características económicas y socioculturales de cuatro aldeas ladinas de Guatemala. Guatemala Indígena, 7:1, 1973.

3. Yarbrough, C., Habicht, J-P., Malina, R.M., Lechtig,
    A. and Klein, R.E.: Length and weight in rural
    Guatemalan Ladino children. Birth to seven years
    of age. Guatemala, Instituto de Nutrición de
    Centro América y Panamá, INCAP Publication No.
    DE-887, 1973.

4. McCammon, R.W.: Human Growth and Development.
    Springfield, Charles C Thomas, 1970.

5. Habicht, J-P., Lechtig, A., Yarbrough, C. and Klein,
    R.E.: The timing of the effect of supplementation
    feeding on the growth of rural preschool children.
    Paper presented at the IX Congreso Internacional
    de Nutrición, México, D.F., September, 1972.

A NEW APPARATUS FOR STEREOMETRY: MOIRÉ CONTOUROGRAPH

Harumi Terada

From the Department of Anatomy, Kitasato
University School of Medicine, Sagamihara,
Japan

In addition to traditional caliper-and-tape anthropometry, other techniques that provide data for the quantitative analysis of body composition have been explored as morphological approaches to the assessment of nutritional status.[1] These techniques should contribute to widening the scope of somatology and its practical usefulness.

A value for body volume is needed for calculations of body density; these allow inferences about body composition. Many methods have been used for volumetry of the living man:

1. The hydrostatic or water displacement method which utilizes the principle of Archimedes.[2-4] This method is relatively simple but cannot be applied to young children or seriously ill patients.

2. The air displacement method in which the physical relationship among ambient air pressure, temperature, volume, water vapor, and mass is utilized to estimate body volume.[5-8] In this method, technical difficulties arise from chamber pressure changes due to respiratory gas exchange, vaporization, and increasing air temperature.

27

3. The helium dilution method, in which a known
    quantity of helium is injected into the air
    space of the occupied subject chamber and den-
    sity is computed from the resultant concentra-
    tion of helium which is proportional to the size
    of the subject.[9-11] This method is not only
    complex and expensive, but also has limitations
    in accuracy such as the precision of the helium
    analyzer and the uncertainties of reading tem-
    perature and relative humidity.

4. The underwater weighing method in which a subject
    with a nose-clip and snorkel mouthpiece is
    weighed on a platform completely immersed in a
    water tank.[12-14] The subjects have to be com-
    pletely accustomed to the procedure before
    measurements are taken, because there is a
    danger that the subjects may drown.

None of the above-mentioned volumetric techniques
is feasible for routine clinical application. Accordingly,
the prediction of body density from other measurements,
such as fatfold, height-weight and girths, has been sub-
stituted.[15-19]

Body surface area has been measured directly using
coating techniques.[20-25] The main requirements are to
affix the tape, flat and smooth, onto the entire surface
of the skin and to remove this mould without altering
the area. This operation requires tremendous time and
patience as well as practice and skill. Accordingly,
it has been customary to use height and weight to calcu-
late the assumed surface area of the body with the aid
of various height-weight formulae or nomograms based on
limited direct measurements. Therefore, in practice,
surface area was not a measurement, but a numerical estimation based on a formula. This is one reason why Durnin[26]
objected to its use as a standard reference for basal
metabolism.

A photographic technique to determine body volume
and surface area is more feasible in surveys of large
numbers of subjects. Several methods of photogrammetry
have been reported: stereophotogrammetry;[27-30] multiplex

projector;[31] photodermoplanimeter;[32-34] cyrtographometer;[35] and monophotogrammetry with colored strips.[36-38]

Stereophotogrammetry is based on photomapping techniques with two cameras and the principle is no different from that of the old polar stereoscope. The contour lines are plotted from the stereo pictures with the aid of a drawing machine (autograph). This method is accurate, but it requires at least two overlapping photographs of the subject, highly skilled personnel and extraordinarily expensive machines to plot the contour lines. The present author was engaged in the three-dimensional study of the human body in cooperation with Institute of Industrial Science, University of Tokyo, but was compelled to abandon this technique in mass surveys because of its expense.

The principle of Burke's[31] method is based on the geometry of the multiplex projector system, which is used for plotting area maps. A life-size visual image is reconstituted. His method reduces the expense, but a skilled specialist is necessary to plot the contour lines. Moreover, the size of the object is limited to the face. It is not practical to apply this method to three-dimensional measurements of the whole human body.

Photometry, using a photodermoplanimeter, determines the radiation surface area of the human body. The radiation surface is ascertained by the reverse method of measuring the subject's absorbing surface when that surface has been made 95 percent absorbent to visible light by means of a pigment. Van Graan[25] has compared the radiation area with the total area measured by coating the body with masking tape, and found the latter to be 4.06 percent greater in 15 Bantu men. This method is simple and inexpensive, but the results do not represent the total surface area of the body. Moreover, this method does not measure body volume.

The cyrtographometer was designed by Roche and Wignall[35] to record mammary gland contours. The principles of the method are the projection of grid shadows with parallel beams of light placing a small light source at the focal point of a spherical mirror. In this method, quantitative analysis is difficult because the contour intervals are too wide and the shaded areas too large.

In the monophotogrammetric method of Pierson,[36] the
human subject is illuminated from either side by lights
that shine through colored transparent strips of equal
width. He is photographed from the front and the back.
The color transparency resulting from photography is
projected onto drafting cloth and an isopleth map is
traced. The volume and surface area are calculated from
lines of demarcation between adjacent colored strips.
This technique seems practical, but the demarcation lines,
which are comparable to contour lines, are frequently
obscured by shadows on shaded areas because of the lateral
sources of illumination. Tracing of the projected image
is time consuming.

Recently Takasaki[39-43] has developed a new system
of moiré interferences of a grid with its shadow cast
onto the surface of an object to visualize contour lines
in situ with a contour interval of 1 to 6 mm. With this
technique, the volume and surface area can be measured
accurately and inexpensively. At the laboratory of the
present author, a moiré apparatus for human stereometry
has been under development since 1970 in cooperation with
Kowa Optical Co.; a provisional apparatus was manufactured
in January 1973. This communication describes the new
device and its utility in providing three-dimensional
data for nutritional study.

THEORETICAL BASIS OF MOIRÉ SYSTEM

The theoretical basis of the moiré system was described in detail by Meadows et al.[44] and Takasaki.[39] In
the present paper, this will be explained on the basis
of geometric optics. The principle of the formation of
moiré contour lines is shown in Figure 1. An equispaced
plane grating with line spacing or pitch of s is illuminated by a point light source L and is observed at E. The
points E and L lie on a plane parallel to the grating.
The distance of E or L from the grating is l. The interval between E and L is d.

Suppose a light beam proceeds from L through A
(center of a line space) perpendicular to the grating.
When the beam is observed from E through B (the center
of the next space), the point A_1 is seen as a bright spot.
When the same beam is observed from E through C (center
of the third space), the point A_2 is a bright spot, and

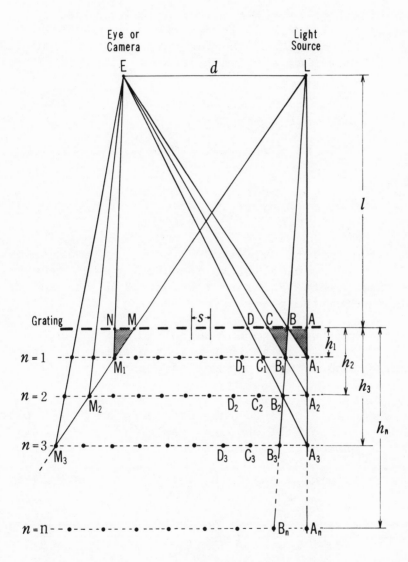

Figure 1.   Theoretical formation of contour lines.

so forth. If a beam proceeds from L through B and is
observed from E through C,D,......, the bright spots will
be seen at B_1, B_2, B_3,.....B_n. Also if another beam proceeds from L through M and is observed from E through
N,....., the bright spots will appear at M_1, M_2, M_3,......,
M_n.

The triangles with two common apical angles, A_1EL
and A_1BA, are similar, and so are the triangles B_1EL
and B_1CB,....., and the triangles M_1EL and M_1NM. According to the theorem of similar triangles,

$$\frac{AB}{LE} = \frac{A_1A}{A_1L}, \quad \frac{BC}{LE} = \frac{B_1B}{B_1L}, \quad \cdots\cdots \quad \frac{MN}{LE} = \frac{M_1M}{M_1L}$$

and $AB = BC = \ldots\ldots = MN = s, \quad EL = d$

Accordingly,

$$\frac{A_1A}{A_1L} = \frac{B_1B}{B_1L} = \cdots\cdots \frac{M_1M}{M_1L} = \frac{s}{d}$$

Thus, a set of points A_1, B_1, C_1, ..., M_1 must be
aligned on a plane parallel to the grating. In other
words, these points will form a bright plane or band of
equal depth from the plane of the grating.

In the same way, another set of points A_2, B_2,
C_2,....., M_2 will form another bright plane parallel to
the grating, and so will the set of points A_n, B_n, ...
M_n, in general.

As shown in Figure 2, the surface of an object placed
behind the grating will show stripes of contour lines,
that can be photographed with a camera at the point E.

When A_1A is h_1, A_2A is h_2,······, A_nA is h_n (Fig. 1),

$$\frac{h_1}{1 + h_1} = \frac{s}{d}, \quad \frac{h_2}{1 + h_2} = \frac{2s}{d}, \quad \cdots\cdots \quad \frac{h_n}{1 + h_n} = \frac{ns}{d}$$

where n is the ordinal number of a contour line, counted
from the plane of the grating.

Accordingly, the depth of the nth contour line (h_n) is obtained as follows:

$$dh_n = nsl + nsh_n$$

$$h_n(d - ns) = nsl$$

$$h_n = \frac{nsl}{d - ns} \qquad \text{for a bright contour line}\ldots\ldots(1)$$

$$\text{or} \quad \frac{(2n - 1)sl}{2d - (2n - 1)s} \qquad \text{for a dark contour line}\ldots\ldots\ldots\ldots(2)$$

And the ordinal number of a contour line (n) is:

$$n = \frac{h_n d}{s(1 + h_n)} \qquad \text{for a bright contour line}\ldots\ldots(3)$$

$$\text{or} \quad \frac{2dh_n - s(h_n - 1)}{2s(1 + h_n)} \qquad \text{for a dark contour line}\ldots\ldots\ldots\ldots(4)$$

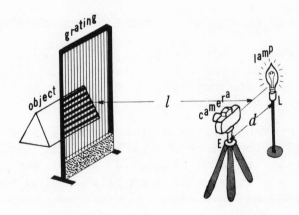

Figure 2. Schematic layout for moiré contourography.

The depth interval between successive dark contour lines or contour interval Δh_n is:

$$\Delta h_n = h_{n+1} - h_n = \frac{4sdl}{(2d - 2ns)^2 - s^2} \quad \cdots\cdots\cdots\cdots (5)$$

The value of d can be calculated from the following formula:

$$d = \frac{s(1 + 2\Delta h_n) \pm \sqrt{s^2(1 + 2\Delta h_n)^2 - \Delta h_n^2 s^2(4n^2 - 1)}}{2\Delta h_n} \quad \cdots (6)$$

In this way, the moiré patterns with the optional value of Δh_n may be actualized, if appropriate values of l, s, and d are chosen in the set up of the apparatus. For example, if Δh_1 is expected to be 4 mm when l is 2.5 m (2,500 mm) and s is 2 mm, the value of d as obtained from the formula (6) is 1,254 mm. In this case, the depth interval between successive contour lines is calculated from the formula (5) and is shown in Table I. The contour interval (Δh_n) enlarges with increasing values of n, when the values of l, d and s are constant.

APPARATUS AND PROCEDURE OF MOIRÉ CONTOUROGRAPHY

The apparatus at our laboratory in Kitasato University was manufactured by the Kowa Optical Co.* and is called "Moiré Contourograph, Type C" (Figures 3 and 4). It consists of the following parts:

1. Two interchangeable gratings composed of black-coated nylon threads arranged vertically: a grating with 1 mm pitch, 700 mm x 1,000 mm in size; a grating with 2 mm pitch, 1,900 mm x 1,000 mm in size.

2. A grating case with a motor that can shuttle

* 3-3-1 Chofugaoka, Chofu-shi, Tokyo, Japan

TABLE   I

Theoretical Values of the
Contour Interval Δh_n, when
N is 2,500 mm, d is 1,254 mm,
and s is 2 mm

n	(between)	Δh_n (mm)
1	(2 - 1)	4.0000
2	(3 - 2)	4.0128
3	(4 - 3)	4.0257
4	(5 - 4)	4.0386
5	(6 - 5)	4.0516
6	(7 - 6)	4.0647
7	(8 - 7)	4.0778
8	(9 - 8)	4.0910
9	(10- 9)	4.1042
10	(11-10)	4.1175
12	(13-12)	4.1444
14	(15-14)	4.1714
16	(17-16)	4.1988
18	(19-18)	4.2264
20	(21-20)	4.2543
22	(23-22)	4.2825
24	(25-24)	4.3110
26	(27-26)	4.3397
28	(29-28)	4.3687
30	(31-30)	4.3980
35	(36-35)	4.4726
40	(41-40)	4.5492
45	(46-45)	4.6277
50	(51-50)	4.7082
55	(56-55)	4.7909
60	(61-60)	4.8758
70	(71-70)	5.0524
80	(81-80)	5.2388
90	(91-90)	5.4357
100	(101-100)	5.6440

the grating in a transverse plane over the distance of 120 mm at a speed of 60 mm/second.

3.   Two xenon short-arc lamps of 1 Kw at 100 V, with lamp-houses and cooling fans.

4.   A cross-rail for lamp-houses, with arm length of 3,300 mm, the range of distance between the camera and lamp (d) being 470 to 1,580 mm.

5.   A camera holder, attached to the cross-rail. The camera and lamps are set precisely on the same plane parallel to the grating.

6.   Two side rails 3,150 mm in length. The range of distance between the camera and the grating (l) is 930 to 3,160 mm.

7.   A control box with remote controlled automatic shutter release and on-off switches for the motor.

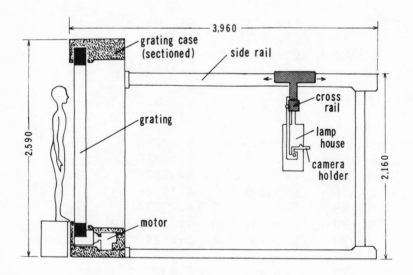

Figure 3. Diagrammatic side view of a moiré contourograph.

    8.   Two lamp ballasts of 100 V, 27A.

    The extent to which the shadow of the grating is
blurred by penumbra is determined by the width of light
source. The smaller or thinner the light source, along
the line of the grating, the sharper the contour line.
The xenon short arc lamp proved to be the best for this
purpose. In order to obtain distinct bright and dark
stripes on the object surface, the room should be shielded
from external light.

    To obtain shadow free illumination, two light sources
are used. Theoretically, it is apparent that the same
moiré patterns are visualized, if the two light sources
are arranged symmetrically against point E (camera position).

    The lateral shuttle movement of the grating is made
during exposure to erase spurious moiré and unnecessary
images of grating-threads from the picture. The spurious
moiré is a pattern formed between the higher harmonics

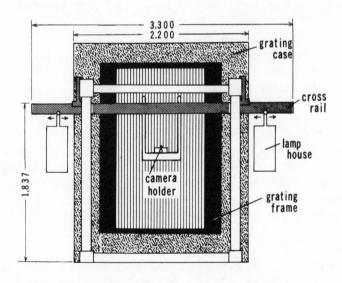

        Figure 4.   Diagrammatic front view of a
                 moiré contourograph.

of the shadow and the grating or between higher harmonics
of the grating and the shadow.

Photography was accomplished by means of a Nikon F-2
camera. To avoid distortion of the image, photography
at a long distance with a lens of long focal length is
desirable. In moiré contourography, however, both the
camera and light sources must be on the same plane parallel to the grating, and lamps with a limited output of
light must be moved from the subject in harmony with movement of the camera. Also the aperture of the camera lens
should be small to increase the depth of focus. A lens
of short focal length allows the use of a larger aperture
than a lens of longer focal length at the sacrifice of
image quality. A shutter speed at exposure faster than
1/8 second is recommended to minimize motion by the subject. As a compromise among these limiting factors, the
exemplar in setting up the apparatus and camera is as
follows:

For whole body: l = 3 m, F 11, 1/8 second with
                 55 mm lens.

For parts of body: l = 2 m, F 16, 1/8 second with
                    200 mm lens.

The film used for this purpose is Kodak Tri-X Pan. The
ASA index is increased to 1,600 with the aid of a special
developer containing phenidon.

The subject under study is requested to stand as
close as possible to the grating because the contour lines
are sharper near the grating. Our apparatus gives visible
moiré patterns of 4 mm interval to a depth of 500 mm using
a grating with 2 mm pitch. Several photographs are taken
of each subject, turning the subject around so that the
anterior, posterior and lateral surfaces show contour
lines. Examples are shown in Figure 5.

The images of points with different distances from
the camera lens appear on the photograph with different
magnification. The formula for the longitudinal magnification is as follows:

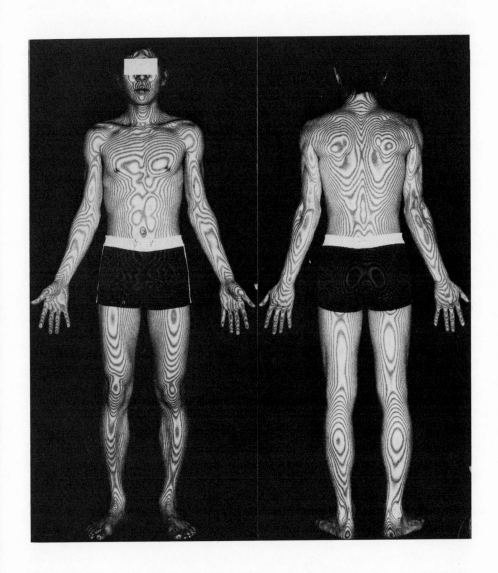

Figure 5. An example of moiré contours. Pitch of the grating, s = 2.0 mm, camera-grating distance, l = 3,000 mm, camera-lamp distance, d = 1,054 mm and contour interval: Δh_1 = 4.0 mm.

$$m_n = \left(\frac{f}{a_n - f}\right)^2$$

where f is a focal length of the lens and a_n is a distance between the lens and a certain point on a contour line at n, and

$$a_n = 1 + h_n$$

The ratio of magnification of Δh_1 to that of Δh_n is

$$\frac{m_1}{m_n} = \frac{(1 + h_n - f)^2}{(1 + h_1 - f)^2}$$

This correction coefficient must be applied to the calculation of Δh_n on the photograph. So far as the contour intervals are concerned, the difference between theoretical values and actual measurements with this apparatus is within one percent.

Three dimensional measurements, such as volume and surface area, are made from the photograph with the aid of a computer. These results will be more meaningful if corrections are made for the volume of air in the lungs and respiratory passages. Studies on the procedures for the input of pattern data, by optic scanning, tracing with a "graph pen" and use of a television camera and monitor, and programming for calculation are in progress. The results will be validated by comparison with those obtained from the methods described earlier.

DISCUSSION

Moiré contourography shows promise as a means of three dimensional analysis of the human body. It has an advantage over other methods in that the subject is little inconvenienced and both surface area and volume can be determined from the same set of data as accurately as expensive and time consuming stereophotogrammetry. The reliability of data obtained from this technique is being

analyzed. The unit cost of material is similar to that
of ordinary monochromatic photography. The time taken
for the accomplishment of photography is not more than
several minutes per subject. Later, a cine camera may
be included to allow a three dimensional analysis of
movements.

Besides volume and surface area, the outline of a
cross-section at any level or in any direction can be
obtained from moiré contour maps (Figure 6). These may
provide data for detailed analysis of nutritional status,
such as swelling or emaciation of particular parts of
the body.

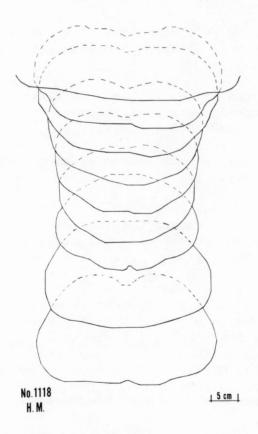

No.1118
H. M.

⊥ 5 cm ⊥

Figure 6. Horizontal sections of the subject
shown in Figure 5.

One limitation of moire contourography, as of other
methods, is the lack of information about internal conditions such as the amount of gas in the gastro-intestinal
tract or of residual air in the lung. A correction for
residual lung volumes can be made using other techniques,
such as the three breath nitrogen dilution method.

REFERENCES

1.   Brožek, J.: Research on body composition and its
     relevance for human biology. In: Brožek, J. (Ed)
     Human Body Composition: Approaches and Applica-
     tions. Proceedings, Symposium Society Study Human
     Biology. Oxford, Pergamon Press, 1965.

2.   Mizuno, T. and Takahashi, K.: A somatological study
     of body capacity of the Japanese. Taiikugaku
     Kenkyu, 4:1, 1960.

3.   Garn, S.M. and Nolan, P., Jr.: A tank to measure
     body volume by water displacement (BOVOTA).
     Ann. N.Y. Acad. Sci., 110:91, 1963.

4.   Jones, P.R.M.: A body volumeter to measure human
     body density. J. Physiol., 222:5, 1972.

5.   Noyons, A.K.M. and Jongbloed, J.: Über die
     Bestimmung des wahren Volumens und des spezifischen
     Gewichtes von Mensch und Tier mit Hilfe von
     Luftdruckveränderung. Pflüger's Arch. ges
     Physiol., 235:588, 1935.

6.   Falkner, F.: An air displacement method of measuring
     body volume in babies: A preliminary communication.
     Ann. N.Y. Acad. Sci., 110:75, 1963.

7.   Lim, T.P.K.: Critical evaluation of the pneumatic
     method for determining body volume: Its history
     and technique. Ann. N.Y. Acad. Sci., 110:72, 1963.

8. Gnaedinger, R.H., Reineke, E.P., Pearson, A.M., Huss, W.D.V., Wessel, J.A. and Montoye, H.J.: Determination of body density by air displacement, helium dilution, and underwater weighing. Ann. N.Y. Acad. Sci., 110:96, 1963.

9. Siri, W.E.: Apparatus for measuring human body volume. Rev. Sci. Instr., 27:729, 1956.

10. Fomon, S.J., Jensen, R.L. and Owen, G.M.: Determination of body volume of infants by a method of helium displacement. Ann. N.Y. Acad. Sci., 110:80, 1963.

11. Clive, D., Ball, M.F., Meloni, C.R., Werdein, E.J., Canary, J.J. and Kyle, L.H.: Modifications of the helium dilution method of measuring human body volume. J. Lab. Clin. Med., 66:841, 1965.

12. Young, C.M., Gehring, B.A., Merrill, S.H. and Kerr, M.E.: Metabolic responses of young women while reducing. J. Am. Diet. Assoc., 36:447, 1960.

13. Durnin, J.V.G.A. and Taylor, A.: Replicability of measurements of density of the human body as determined by underwater weighing. J. Appl. Physiol., 15:142, 1960.

14. Katch, F., Michael, E.D. and Horvath, S.M.: Estimation of body volume by underwater weighing: Description of a simple method. J. Appl. Physiol., 23:811, 1967.

15. Young, C.M., Martin, M.E., Tensuan, R. and Blondin, J.: Predicting specific gravity and body fatness in young women. J. Am. Diet. Assoc., 40:102, 1962.

16. Sloan, A.W., Burt, J.J. and Blyth, C.S.: Estimation of body fat in young women. J. Appl. Physiol., 17:967, 1962.

17.   Durnin, J.V.G.A. and Rahaman, M.M.: The assessment
      of the amount of fat in the human body from
      measurements of skin fold thickness. Brit. J.
      Nutr., 21:681, 1967.

18.   Katch, F.I. and Michael, E.D., Jr.: Prediction of
      body density from skin-fold and girth measurements
      of college females. J. Appl. Physiol., 25:92,
      1968.

19.   Michael, E.D., Jr. and Katch, F.I.: Prediction of
      body density from skin-fold and girth measurements
      of 17-year-old boys. J. Appl. Physiol., 25:747,
      1968.

20.   DuBois, D. and DuBois, E.F.: A formula to estimate
      the approximate surface area if height and weight
      be known. Arch. Int. Med., 17:863, 1916.

21.   DuBois, D. and DuBois, E.F.: A height-weight
      formula to estimate surface area of man. Proc.
      Soc. Exp. Biol., 13:77, 1916.

22.   Takahira, H.: The basal metabolism of normal
      Japanese men and women. Eiyokenkyujo Hokoku,
      1:61, 1925.

23.   Murata, T.: Studies on the body surface area of
      the Japanese female. I, II, III, IV. Shikoku
      Igaku Zasshi, 15:495, 1959.

24.   Sendroy, J., Jr. and Collison, H.A.: Nomogram for
      determination of human body surface area from
      height and weight. J. Appl. Physiol., 15:958,
      1960.

25.   Van Graan, C.H.: The determination of body surface
      area. South African Med. J., 43:952, 1969.

26.   Durnin, J.V.G.A.: Somatic standards of reference.
      In: Brožek, J. (Ed) Human Body Composition:
      Approaches and Applications. Proceedings,
      Symposium Society Study Human Biology. Oxford,
      Pergamon Press, 1965.

27. Hertzberg, H.T.E., Dupertuis, C.W. and Emanuel, I.:
     Stereophotogrammetry as an anthropometric tool.
     Photogram. Eng., 23:942, 1957.

28. Pierson, W.R.: Non-topographic photogrammetry as
     a research technique. FIEP Bull., 27:48, 1957.

29. Haga, M., Ukiya, M., Koshihara, Y. and Ota, Y.:
     Stereophotogrammetric study on the face. Bull.
     Tokyo Dent. Coll., 5:10, 1964.

30. Maruyasu, T., Oshima, T., Yanagisawa, S., Hasebe,
     Y. and Matsuyama, Y.: Measurement of human body
     by stereophotogrammetry. I. Ningen Kogaku,
     4:258, 1967.

31. Burke, P.H.: Stereophotogrammetric measurement of
     normal facial asymmetry in children. Human Biol.,
     43:536, 1971.

32. Halliday, E.C. and Hugo, T.J.: The photodermoplani-
     meter. J. Appl. Physiol., 18:1285, 1963.

33. Van Graan, C.H. and Wyndham, C.H.: Body surface
     area in human beings. Nature, 204:998, 1964.

34. Mitchell, D., Strydom, N.B., Van Graan, C.H. and
     Van Der Walt, W.H.: Human surface area: Compari-
     sons of the DuBois formula with direct photometric
     measurement. Pflüger's Arch., 325:188, 1971.

35. Roche, A.F. and Wignall, J.W.G.: The cyrtograph-
     ometer: A new instrument for recording contours.
     Am. J. Phys. Anthrop., 20:521, 1962.

36. Pierson, W.R.: Monophotogrammetric determination
     of body volume. Ergonomics, 4:213, 1961.

37. Pierson, W.R.: The estimation of body surface area
     by monophotogrammetry. Am. J. Phys. Anthrop.,
     20:399, 1962.

38. Pierson, W.R.: A photogrammetric technique for the
    estimation of surface area and volume. Ann. N.Y.
    Acad. Sci., 110:109, 1963.

39. Takasaki, H.: Moiré topography. Applied Optics,
    9:1457, 1970.

40. Takasaki, H.: Moiré topography. Shashin Sokuryo,
    10:1, 1971.

41. Takasaki, H.: Moiré topography. Gazo Gijitsu, II:
    34, 1972.

42. Takasaki, H.: Moiré topography. Keisoku to
    Seigyo, 12:390, 1973.

43. Takasaki, H.: Moiré topography. Applied Optics,
    12:845, 1973.

44. Meadows, D.M., Johnson, W.O. and Allen, J.B.:
    Generation of surface contours by moiré pattern.
    Applied Optics, 9:942, 1970.

OVERNUTRITION, OVERGROWTH AND HORMONES (WITH SPECIAL

REFERENCE TO THE INFANT BORN OF THE DIABETIC MOTHER)*

Donald B. Cheek, James B. Brayton, and Rachel
E. Scott

From the Research Foundation, Royal Children's
Hospital, Parkville, Melbourne, Australia, and
the Department of Pediatrics, and the Children's
Medical and Surgical Center, the Johns Hopkins
Hospital, Baltimore, Maryland, USA

It is the purpose of this paper to consider the
effects of overnutrition on changes in body composition
and cell growth and to discuss such changes in relationship to hormone secretion. The role of insulin secretion
on cell growth is of importance especially with respect
to cell growth in the primate fetus.

The study of growth extends from the earliest phase
of embryological development to the last stage of senility
when metabolically active cells are lost. The latter can
be demonstrated easily from existing studies with K^{40}
counting.[1] In our work on growth and body composition,
we have been impressed with the need to assess body size,
metabolic function and maturational age--whether one con-
siders the body as a whole, an individual tissue, or the
growth of cells.[1] Genetic and environmental factors are
of the greatest importance. However, the role of environment has not been appreciated fully in terms of nutrition

* This work was supported by grant HD 00126-08 from the
  National Institute of Child Health and Human Development,
  National Institutes of Health, Bethesda, Maryland, USA.

and energy expenditure. Social customs within the
environment may have far-reaching effects on somatic
growth, and would appear to be definite factors in
longevity.[2]

In any country where longevity is found and where
individuals live to 130 years, the older people participate in continued exercise and have a reduced calorie
and animal fat intake,[2] all of which is in striking contrast to urban life in affluent countries of today. In
experimental animals, continued exercise ensures growth
of new muscle cells and accretion of deoxyribonucleic
acid (DNA), ribonucleic acid (RNA) and protein.[3,4]
Muscle mass increases and the cellular components increase commensurately. Provided exercise is not too
strenuous, appetite does not increase[5] and a correct balance is kept between lean and adipose tissues.

The role of hormones in the induction of growth is
intimately related to nutrition because various substrates
stimulate hormone production. Attention has been drawn
to the sequence of events following food ingestion and
to the secretion of insulin and growth hormone.[6,7] Periods
of fasting and feasting can be related to phases of hormone release, in particular the release of insulin or
growth hormone.

Mechanisms involved in normal growth and in excess
somatic growth, be they fetal, neonatal or adolescent,
are the subject of research.[8] The fact that many infants
and high school students in the United States are now
obese or overweight for age is a matter for concern.[9,10]
The cessation of breast feeding and departure from marked
physical activity, which previously was inherent in rural
life, are but two examples of changes in life patterns.
High density feeding in the neonatal period leads to excessive somatic and skeletal growth.[11,12] Similar conclusions have been derived from studies of infant baboons
given an unlimited diet when confined to a cage.[13]

In an attempt to obtain information concerning the
role of hormones and nutrition in fetal growth, we have
studied the Rhesus monkey (Macaca mulata).[14-21] The use
of human fetal material would involve limitations. The

use of a subhuman primate is advantageous because primate growth differs significantly from non-primate mammalian growth.[21,22]

FETAL OVERGROWTH

Fetal overgrowth occurs in the infant born of a diabetic mother (subsequently referred to as IBDM).[23,24] Excess adipose tissue is found coincidentally with excessive lean body mass and visceral enlargement has been recorded.[25] Various theories concerning this excessive growth have ranged from the possible role of fetal growth hormone,[26,27] to excessive activity of the adrenal gland,[28,29] to hyperplasia, hypertrophy, and excess activity of the islet cells (or β cells) within the pancreas.[30,31] The latter would appear correct. The reason for the excess release of fetal pancreatic insulin in IBDM can be ascribed to fluctuations in maternal and fetal glucose together with increased responsiveness to amino acid levels within the fetal circulation.[32]

Mild diabetes mellitus in the pregnant Rhesus monkey has been induced by Mintz et al.[32] They showed that moderate diabetes with mild ketosis was associated with depressed plasma insulin and decreased glucose tolerance. The fetus was oversized, with a large placenta, and resembled the human infant born of the diabetic mother (IBDM). There was a reduction in maternal plasma insulin and decreased glucose tolerance. These findings are in agreement with our own experience.

In mid-gestation, we gave pregnant primates (Macaca mulata) a single dose of streptozotocin intravenously-- this drug ablates the pancreatic β cells. Nine abnormal fetuses were delivered normally or taken by serial section between 143 and 155 days. These nine fetuses were compared with 22 normal fetuses delivered by Cesarian section from 140 to 160 days. Full term for the fetal Macaca mulata is 165 days.

In this work the cerebrum, cerebellum, muscle and other tissues (where appropriate) were analyzed separately for DNA, RNA, protein, zinc, carbonic anhydrase activity, lipid N-acetyl neuraminic acid (lipid NANA) and chloride.[17] Organ weights and fat content were analyzed statistically, using the student's t test, while the appraisal of other chemical determinants was made by analyses of covariance.

Somatic Changes

The fetuses or infants born of diabetic mothers (IBDM) were variable in weight but five out of nine were greater than one standard deviation above the control values while four were two standard deviations or more above. Five had a disproportionate increase in length. The carcass (muscle plus skeleton) was excessive in some and likewise the visceral mass. Some organs were affected more than others (heart, spleen, pituitary and adrenals; Table I). The amount of fat in the carcass was increased ($p < 0.02$), likewise the percent fat in muscle ($p < 0.001$) and skin ($p < 0.005$).

Carcass mass (muscle plus skeleton) showed an increase ($p < 0.02$). The DNA content (number of nuclei within the muscle mass) appeared to be decreased but the DNA content per gram of muscle was decreased ($p < 0.01$). The significant change in muscle was the increase in the ratio of protein to DNA ($p < 0.01$) indicating an advancement in cytoplasmic growth (Figure 1). This change was emphasized by the finding that the percent water in muscle was reduced ($p < 0.001$). The total protein in the muscle was increased ($p < 0.025$).

Brain Changes

Biochemical changes were found in the brain. The cerebrum was represented by that part of the brain remaining after section between the inferior and superior colliculi and after removal of the cerebellum at the peduncles. The weight of the cerebrum was normal for age. The DNA content (or cell number) was less than normal ($p < 0.01$). The total protein of the cerebrum tended to be increased and the protein to DNA ratio was markedly elevated ($p < 0.01$) in all instances (Figure 2). The Zn:DNA ratio was increased similarly. The RNA:DNA ratio showed a trend

TABLE I

Organ Weights and Percent Fat

	Pituitary (mg)	Heart (g)	Spleen (g)	Adrenals (mg)	Carcass fat (g)	% Fat Muscle	Skin
Control							
Mean	13.3	2.09	0.57	240	5.39	1.03	5.59
s.d.	2.1	0.22	0.17	81	1.95	0.42	2.48
N	19	22	18	22	22	22	11
I.B.D.M.							
Mean	16.0	2.34	0.79	330	10.12	2.85	12.80
s.d.	1.2	0.32	0.18	96	5.42	1.30	5.99
N	9	9	9	9	9	9	9
t	4.31	2.13	2.95	2.46	2.55	4.10	3.38
P	<0.001	<0.05	<0.01	=0.02	<0.02	<0.001	<0.005

towards an increase (five out of nine points > one s.d.
above the mean) while the percentage of water was decreased
(p < 0.025). The activity of carbonic anhydrase also
showed a trend towards an increase (five out of nine
points > one s.d. above the mean).

The cerebellum was normal in weight but its carbonic
anhydrase activity was increased significantly (p < 0.001)
while the concentration of water was decreased (p < 0.01).

Discussion

These findings, presented for the first time in the
literature, confirm and extend what has been found in the
human fetus with overgrowth consequent to maternal diabetes. While the normal fetal monkey has very little
adipose tissue at birth, the presence of fat was obvious
in these fetuses. The enlargement of the spleen was of

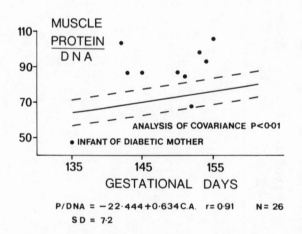

P/DNA = −22·444 + 0·634 C.A.   r = 0·91   N = 26
SD = 7·2

Figure 1. The ratio Protein: DNA is increased in the
muscle of infants born of diabetic mothers (Macaca
mulata). This ratio indicates an increased cytoplasmic
growth within the muscle fibre. The lines (—— and ---)
indicate the mean ± 1 s.d. for normal infants. The regression and other parameters refer to normal infants.

interest insofar as the high hematocrit of these infants had been ascribed by us to reduced blood volume.[24] The size of the spleen and the hemopoietic tissue is known to be increased[33] but the extracellular volume is reduced.[24]

The finding that the pituitary and adrenal glands are enlarged raises interesting questions. It could be hypothesized the secretion of several trophic hormones is increased. Perhaps the discharge of growth hormone, adrenocortical hormone and other hormones are all increased as a result of hypothalamic stimulation. Testicular inter-stitial cell hyperplasia and the formation of luteinized ovarian cysts have been documented in the human infants born of diabetic mothers.[33]

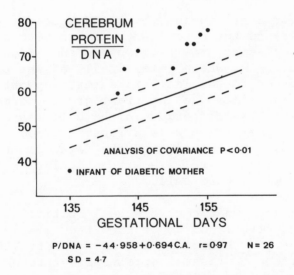

P/DNA = −44·958+0·694C.A.   r= 0·97   N = 26
SD = 4·7

Figure 2. The ratio Protein: DNA is shown for the cerebrum of infants born of diabetic mothers (Macaca mulata). The data demonstrate increased cytoplasmic growth in the brain cells. The lines (—— and ---) indicate the mean ± 1 s.d. for normal infants. The regression and other parameters refer to normal infants.

The mechanism for increased insulin release might
be related also to hypothalamic activity because βcell
activity is modulated by hypothalamic cells.[34,35] The
over-riding role of insulin in this situation is supported
by findings in muscle. Our work on postnatal weanling
rats, over the last 10 years, supports the thesis that
insulin is responsible for cytoplasmic growth in muscle
while DNA replication is related directly to growth
hormone.[3,36] From our work and that of others, it appears
that growth hormone causes the release of a peptide
somatomedin which is thought to be responsible for cell
multiplication, although it competes with insulin at the
membrane level.[37]

In the absence of the pituitary, insulin can induce
significant cytoplasmic growth in muscle. In our studies
on hypophysectonized rats, it was clear that the continued growth of muscle was related to cytoplasmic growth,
not to nuclear number (or more DNA units), and that it
could be prevented by streptozotocin injection.[36]

From the present studies, it would appear that insulin can influence the cytoplasmic growth of brain cells
as well as cytoplasmic growth in muscle and in other tissues. Previous work, where fetal βcells of the pancreas
were ablated prior to the last trimester, supports our
attitude that insulin is an important growth hormone for
the fetus.[18] Protein accretion in the cell was influenced
and other endocrine organs were affected. Indeed, it
appeared that partial ablation of the fetal pancreas
stimulated reactive hyperplasia with stimulation of protein accretion within the cell.

We do not know what controls cell multiplication in
the soma of the fetus, nor do we know the time in postnatal life when growth hormone begins to exert its cell
multiplying effect on tissues (presumably through somatomedin), or whether somatomedin per se causes cell multiplication in the fetus. A pituitary dwarf is not recognized until some months after birth.[38] Usually the Snell
Smith dwarf mouse with pituitary insufficiency is not
recognized until the 10th postnatal day when weight gain
ceases.[39] The fact that growth hormone can restore most
of the brain changes due to thyroid insufficiency,[40,41]

because acidophil cells degenerate, indicates that growth
hormone might influence fetal brain growth.[17] Work by
Zamerhoff and other investigators[36] suggests that, in
fact, such an influence occurs on the growth of neurones
in the fetal rat. Decapitation in mid-gestation causes
immaturity of the endocrine organs in Macaca mulata and
delayed somatic growth.[42]

It can be mentioned in passing that the increase in
Zn, relative to DNA, in the cerebrum emphasizes the increased protein accretion.[43-45] The increased activity
of carbonic anhydrase (present exclusively in neuroglial
cells[46,47] of the human brain) suggests advancement in
brain development. However, other indices suggest retardation in the human,[48] while our chemical analyses demonstrated a reduced phospholipid content in the macaque
cerebrum. Thus, brain maturation is advanced in some
respects but not others.

The weight of the lung was reduced in most fetuses.
This is of interest because the respiratory distress
syndrome is present commonly in IBDM.

In summary, the excess substrate (glucose and amino
acid) in the fetal circulation may activate the fetal
endocrine system (hypothalamic-pituitary-adrenal axis)
giving rise to excess cytoplasmic growth in lean tissues
and excess adipose tissue growth. Such excess soft tissue
growth can be detected relative to body length well into
the postnatal period.[49]

OBESITY AND OVERGROWTH DURING POSTNATAL LIFE

By six months of age, the normal infant has a high
degree of physiological fatness; 50 percent of the body
weight is adipose tissue.[11] There must be a remarkable
growth of fat cells, either in number or in size or both,
during infancy. There is a paucity of data. While birth
weight has not altered during the last 100 years,[50] the
period at which birth weight doubles seems to be closer
to three months than to the expected six months.
Lightwood (personal communication), in a survey throughout
parts of the United States, found that calorie intake

was 200–300 percent above requirements for infants in "well baby clinics."

Fomon[51] points out that for the United States in the 1950's, 30 percent of infants were breast fed at the time they left the maternity hospital. Today the corresponding figure is 15 percent. Twenty years ago, evaporated milk formula was the routine but today 80 percent of infants receive commercially prepared formulas. After three months of age, the percentage of infants that receive whole cow's milk increases rapidly. By four months of age, less than 10 percent of infants are breast fed. Infants at that age receiving commercial formulas showed an excessive growth rate (length and weight). Fomon points out one advantage of breast feeding. The mother does not know the quantity of food being taken by the infant. Therefore, she does not "feed to the last drop." She knows only when the breast fed infant is satisfied.

When a girl begins to develop sexually, after about nine years of age, significant growth of adipose tissue occurs.[52] Muscle mass begins to double in boys at about 11 years and by sexual maturation, at 14 years, the number of nuclei within the muscle mass has almost doubled. Gain in fat is not remarkable in boys during pubescence. Indeed, some fat may be lost. By contrast, a girl does not show such a dramatic gain in muscle mass or in DNA within her muscle mass.[3] Sex differences are defined clearly with respect to the growth of muscle cells but sex differences in the growth of fat cells have not been demonstrated.

Changes in Lean Tissue and the Role of Hormones

Obese individuals have, we find, excessive growth of lean tissues for age or body length as well as excessive growth of adipose tissue.[53,54] Frequently skeletal growth or body length is excessive for age. Higher fasting insulin levels exist. We have shown that the longer the duration of obesity from birth, the higher the fasting level of insulin.[53] Arginine infusion causes an excess release of insulin[53] and depressed levels of growth hormone in obese children, similar to what one finds in

adults with obesity.[54] However, an insensitivity to
insulin, at the tissue level, appears to develop as
obesity progresses. This is demonstrated by a failure
of labelled glucose or amino acid to be adequately transported into the muscle cell, as well as changes in the
response of the adipocyte.[55,56] The plasma levels of
growth hormone move in the opposite direction to those
of insulin. Whether the changes in concentration are
due to changed uptake by the tissues or changed secretion
is not clear.

Almost all obese adolescent males and females have
an increased lean body mass for age or length, whether
they have advanced bone age or not. Our data do not support the earlier work of Forbes[57] where K^{40} was used to
assess the lean body mass. The difference may be due,
in part, to the fact that K^{40} does not monitor lean body
mass accurately in obesity because the adipose tissue
acts as a shield.[54] One could be concerned[58] that, even
during normal adolescence in the female, some data obtained from K^{40} counting may be incorrect due to the
physiological deposition of fat. Flynn et al.[59] have
shown sex differences in the growth of cell mass, relative
to length, at about 135 cm in height (10 years). In previous studies, we found that the linear relationships
between body water and body potassium differed for boys
and girls.[60]

The increase in lean tissue in obesity is in excess
of that expected, using body length as a base line. A
further assessment can be made by subtracting the extracellular volume (corrected bromide space) from the total
body water yielding intracellular water. This parameter
represents the intracellular water of muscle and visceral
tissues. Because the ratio of water to protein is constant in these tissues, we are in fact measuring the intracellular mass, that is, mainly the metabolically active
tissue (in contrast, for example, to collagen). The intracellular mass is excessive for body length or for age in
obese adolescents[54] (Figure 3).

Muscle mass was excessive for body length in 12
obese adolescent females. These girls had an excessive
amount of DNA in their musculature, relative to length,

indicating an excess of nuclei or DNA units. Their musculature resembled that of the male. If the number of
nuclei within the musculature is considered against chronological age, it can be shown (Figure 4) that the points
for males fall along the expected pathway. However, in
obese girls, especially if there is advanced skeletal
maturation, the points fall well above the expected line.
The departure from the expected values is gross. It is

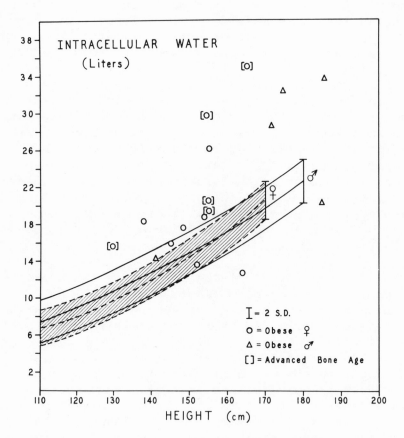

Figure 3. Intracellular water (intracellular mass)
plotted against height for obese adolescent subjects.
The data for normal children is shown as means ± 2 s.d.
Note the increase of intracellular mass in many of the
obese subjects.

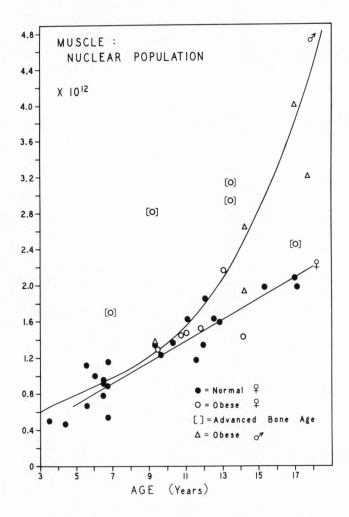

Figure 4.   The muscle nuclear population (DNA content)
plotted against age for normal and obese males and
females.   The expected lines for normal children are
shown also.   The data for obese girls indicate a gross
increase in the number of muscle nuclei, particularly
if bone age is advanced.

reasonable to conclude that the growth in cell number is
due to an excessive response or to an excessive secretion
of growth hormone. However, because estrogens antagonize
growth hormone and/or somatomedin,[61] while androgens
appear to act in the opposite direction,[36] it could be
suggested that androgens predominate in these obese adolescent and pre-adolescent girls. There is some evidence
that this occurs.[62]

By contrast, the response of muscle to insulin
appears to be reduced in established obesity.[55] As one
would predict, the protein:DNA ratio is significantly
reduced. One patient we have studied over a five-year
period provided data relevant to whether this change is
reversible. After nearly two years, he was subjected to
an ileal bypass and studied for a further three years.
As a result of the bypass, his weight decreased from
244.6 kg to 108.7 kg. The reductions in lean body mass,
body fat and body weight are shown in Figure 5. Figure 6
illustrates the loss of nuclear number in the muscle mass
and the reduction of muscle mass. Although the protein:DNA
ratio in the muscle remained unchanged or at significantly
reduced levels, the lean body mass and intracellular mass
returned to expected levels, for height.

OVERGROWTH AND ADIPOSE TISSUE

Finally, one may consider briefly what happens to
the adipose tissue mass in obesity due to overnutrition.
Data on body water (D_2O determinations) and body length
(height age) from infancy to 30 years have allowed the
construction of quadratic equations to predict body fat
for length.[52] There is no physiological relationship
between these measurements but one can still use length
to estimate the expected fat for an individual. For this
purpose, length is a better index of development than
chronological age. The measurement of lean body mass
(from the distribution of D_2O) allows the prediction of
actual fat. "Excess fat" can be appraised as the difference between actual fat and expected fat. The level
of circulating insulin (plasma insulin concentration
multiplied by extracellular volume) is linearly related
in a strongly positive way to the amount of excess fat.[53]

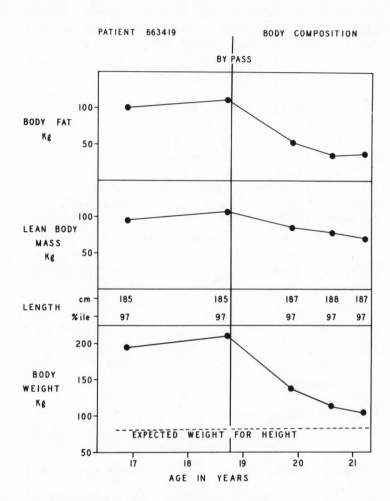

Figure 5. Changes in a grossly obese boy first seen at 17 years of age. When nearly 19 years, he underwent an ileal bypass. In the succeeding 2.5 years, there was a considerable loss of body fat, lean body mass and body weight.

The growth of fat cell number, within the adipose tissue in early infancy, may be a key to the induction of obesity.[63] Knittle and Hirsch[63] estimate that the obese adolescent may have three to four times the normal number of adipocytes. Obese males have a 400 percent increase in collagen within their adipose tissue, a finding that has serious implications for cardiovascular function.

Analyses of small samples of adipose tissue for water, fat, collagen and non-collagen protein yield information concerning fat cell numbers.[64] If the cytoplasmic protein within the adipocyte remains constant, then the percent of non-collagen protein present in a sample should reflect the number of cells per gram of adipose tissue.[54] This appears correct.[64,65] By such an approach, it is possible to demonstrate whether an excess number of adipocytes is present in the adipose tissue of boys and girls.[63]

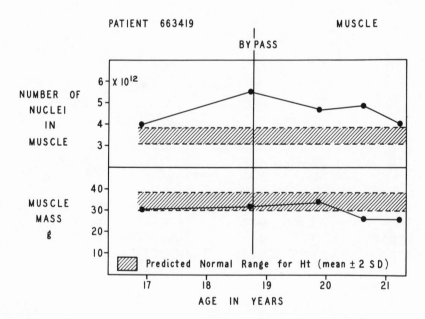

Figure 6. Data from the same patient referred to in Figure 5.

It has been considered that, after full growth has been reached in the individual, the fat cell number reaches a constant level. A similar situation is thought to pertain in obesity. Thus, loss of fatness in the obese individual would occur only through loss of the lipid from adipocytes. This thesis may not always be correct. The patient referred to previously lost some 70 kg of body fat after the ileal bypass, and his fat cell number decreased from 16×10^{10} to 7.5×10^{10} (Figure 7). Admittedly, this patient might be exceptional.

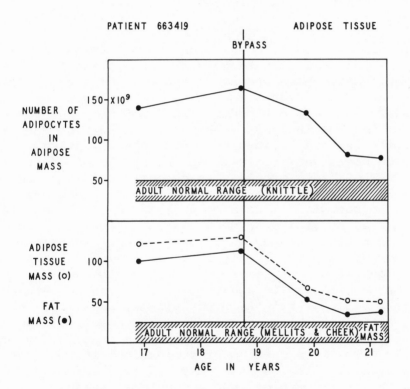

Figure 7. Data from the patient referred to in Figures 5 and 6. There was a loss of about half the adipocytes. This is contrary to the findings of other workers and may be due to a gross deprivation of nutrients causing a strongly negative nitrogen balance.

SUMMARY AND IMPLICATIONS

Work on diabetic primates[32] suggests that the fetal islet cells are very responsive to excess circulating substrate (glucose and amino acids). Excess insulin secretion is accompanied by increased growth of lean and adipose tissue. Increments in cytoplasmic growth are characteristic in muscle and the cerebrum. The pituitary and adrenal glands are involved also but their inter-relationships with the pancreas are not clear--presumably hypothalamic activity is enhanced. Clearly, excess sub-strate and insulin secretion work together to produce accelerated growth and accelerated development of some functions, e.g., brain carbonic anhydrase activity. By contrast, cell multiplication would not appear to be advanced.

The close association between overnutrition, over-growth and hormonal secretion is apparent again from the investigation of obese humans during pre-adolescence or adolescence.[66,67] High circulating plasma levels of insulin are related to the duration of obesity and to the degree of "excess fatness." There is evidence of exces-sive growth in lean tissues and cell multiplication (nuclear number) was excessive in the muscle of obese females. They had a muscle mass on nuclear number com-parable with that of males, whether length or age was used as the base line. Because nuclear number in muscle is influenced by growth hormone secretion, which in turn is said to be enhanced by androgens, adrenal androgens[62] may be involved; particularly since estrogens inhibit some actions of growth hormone.[3] The advancement of bone age in these girls again suggests a relationship between nutrition, cellular growth, hypothalamic, pancreatic and adrenal activity. Also, the findings in muscle tissue indicate that a relative insensitivity to insulin develops gradually. The introduction of serious nutritional re-striction, e.g., ileal bypass, may, to some extent, reverse the overgrowth of lean and adipose tissue.

Changes in tradition and custom, for example, the cessation of breast feeding, predispose to the development of overnutrition in infancy, and, if the infant is geneti-cally prone, this overgrowth may advance to obesity.

Because advancement in maturation is related to early sexual maturation, which in turn is related to life span, one can anticipate that the latter will be reduced and that negative growth, when cells are lost, will occur earlier.[1]

REFERENCES

1. Cheek, D.B.: Conclusions and future implications, p. 616. In: Cheek, D.B. (Ed) Human Growth. Philadelphia, Lea and Febiger, 1968.

2. Leaf, A.: Every day is a gift when you are over 100. Nat. Geog. Mag., 143:93, 1973.

3. Cheek, D.B., Holt, A.B., Hill, D.E. and Talbert, J.L.: Skeletal muscle cell mass and growth: the concept of the deoxyribonucleic acid unit. Pediat. Res., 5:312, 1971.

4. Christensen, D.A. and Crampton, E.W.: Effects of exercise and diet on nitrogenous constituents in several tissues of adult rats. J. Nutr., 86:369, 1965.

5. Cooper, K.H., (Ed): The New Aerobics. New York, M. Evans Company, 1970.

6. Zierler, K.L. and Rabinowitz, D.: Roles of insulin and growth hormone, based on studies of forearm metabolism in man. Medicine, 42:385, 1963.

7. Rabinowitz, D. and Merimee, T.J.: Peripheral actions and regulation of insulin and growth hormone secretion in intact man, p. 207. In: Cheek, D.B. (Ed) Human Growth. Philadelphia, Lea and Febiger, 1968.

8. Cheek, D.B., Graystone, J.E. and Read, M.S.: Cellular growth, nutrition and development. Pediatrics, 45:315, 1970.

9.   Huenemann, R.L., Hampton, M.C., Shapiro, L.R. and
      Behnke, A.R.: Adolescent food practices associ-
      ated with obesity. Fed. Proc., 25:4, 1966.

10.   Mayer, J. and Thomas, D.W.: Regulation of food
      intake and obesity. Science, 156:328, 1967.

11.   Fomon, S.J.: Body composition of the male reference
      infant during the first year of life. Pediatrics,
      40:863, 1967.

12.   Fomon, S.J., Filer, L.J., Jr., Thomas, L.N., Rogers,
      R.R. and Proksch, A.M.: Relationship between
      formula concentration and rate of growth of normal
      infants. J. Nutr., 98:241, 1969.

13.   Buss, D.H. and Voss, W.R.: Evaluation of four
      methods for estimating the milk yield of baboons.
      J. Nutr., 101:901, 1971.

14.   Cheek, D.B.: Protein restriction and brain develop-
      ment. In: Nyhan, W.L. (Ed) Inborn Errors of
      Amino Acid Metabolism. New York, John Wiley and
      Son, 1973.

15.   Myers, R.E., Hill, D.E., Holt, A.B., Scott, R.E.,
      Mellits, E.D. and Cheek, D.B.: Fetal growth
      retardation produced by experimental placental
      insufficiency in the rhesus monkey. I. Body
      weight, organ size. Biol. Neonate, 18:379, 1971.

16.   Hill, D.E., Myers, R.E., Holt, A.B., Scott, R.E.
      and Cheek, D.B.: Fetal growth retardation pro-
      duced by experimental placental insufficiency in
      the rhesus monkey. II. Chemical composition of
      the brain, liver, muscle and carcass. Biol.
      Neonate, 19:68, 1971.

17.   Holt, A.B., Kerr, G.R. and Cheek, D.B.: Prenatal
      hypothyroidism and brain composition in a primate.
      Nature, 243:413, 1973.

18. Hill, D.E. and Cheek, D.B.: Studies on the cellular growth and substrate availability in the rhesus monkey fetus. XIII International Congress of Pediatrics, Vienna, Austria, 1971.

19. Cheek, D.B.: Brain growth and nucleic acids: The effect of nutritional deprivation. Paper presented at the opening of the Mailman-Kennedy Institute, Miami, Florida, March 1971.

20. Cheek, D.B., Holt, A.B. and Mellits, E.D.: Malnutrition and the nervous system, p. 3. In: Nutrition, The Nervous System and Behavior. Pan American Health Organisation Publication #251, 1972.

21. Cheek, D.B., (Ed): Fetal and Postnatal Growth in the Primate: The role of Hormones and Nutrition. New York, John Wiley & Sons (in preparation).

22. Dawes, G.S.: The placenta and foetal growth. In: Dawes, G.S. (Ed) Foetal and Neonatal Physiology. Chicago, Year Book Medical Publishers, Inc., 1968.

23. Osler, M. and Pedersen, J.: The body composition of newborn infants of diabetic mothers. Pediatrics, 26:985, 1960.

24. Cheek, D.B., Maddison, T.G., Malinek, M. and Coldbeck, J.H.: Further observations on the corrected bromide space of the neonate and investigation of water and electrolyte status in infants born in diabetic mothers. Pediatrics, 28:861, 1961.

25. Fee, B.A. and Weil, W.B., Jr.: The body composition of infants of diabetic mothers by direct analysis. Ann. N.Y. Acad. Sci., 110:869, 1963.

26. Westphal, O.: Human growth hormone, a methodological and clinical study. Acta Paediat. Scand., Suppl. 182, 1968.

27. Gaunt, W.D., Bahn, R.C. and Hayles, A.B.: A quantitative cytologic study of the anterior hypophysis of infants born of diabetic mothers. Proc. Mayo Clin., 37:345, 1962.

28. Naeye, R.L.: Infants of diabetic mothers. A quantitative, morphologic study. Pediatrics, 35:980, 1965.

29. Naeye, R.L.: New observations in erythroblastosis fetalis. J.A.M.A., 200:281, 1967.

30. Hoet, J.J.: Normal and abnormal foetal weight gain, p. 186. In: Wolstenholme, G.E.W. and O'Connor, M. (Eds) Fetal Autonomy, A Ciba Foundation Symposium, London, Churchill, 1969.

31. Isles, T.E., Dickson, M. and Farquhar, J.W.: Glucose tolerance and plasma insulin in newborn infants of normal and diabetic mothers. Pediat. Res., 2:198, 1968.

32. Mintz, D.H., Chez, R.A. and Hutchinson, D.L.: Subhuman primate pregnancy complicated by streptozotocin-induced diabetes mellitus. J. Clin. Invest., 51:837, 1972.

33. Driscoll, S.G., Benirschke, K. and Curtis, G.W.: Neonatal deaths among infants of diabetic mothers. Postmortem findings in ninety-five infants. Am. J. Dis. Child., 100:818, 1960.

34. Han, P.W., Yu, Y-K. and Chow, S.L.: Enlarged pancreatic islets of tube-fed hypophysectomized rats bearing hypothalamic lesions. Amer. J. Physiol., 218:769, 1970.

35. Idahl, L.-Å. and Martin, J.M.: Stimulation of insulin release by a ventro-lateral-hypothalamic factor. J. Endocr., 51:601, 1971.

36. Cheek, D.B. and Hill, D.E.: The effect of growth hormone on cell and somatic growth. Handbook of Physiology, Section 7: Endocrinology, in press.

37. Van Wyk, J.J., Hintz, R.L., Clemmons, D.R., Hall, K. and Uthna, K.: Human somatomedin, the growth hormone dependent sulfation and thymidine factor. 4th International Congress of Endocrinology, Washington, D.C., 1972.

38. Brasel, J.A., Wright, J.C., Wilkins, L. and Blizzard, R.M.: An evaluation of seventy-five patients with hypopituitarism beginning in childhood. Amer. J. Med., 38:484, 1965.

39. Cheek, D.B., Powell, K. and Scott, R.E.: Growth of muscle cells (size and number) and liver DNA in rats and Snell Smith mice with insufficient pituitary, thyroid, or testicular function. Bull. Hopkins Hosp., 117:306, 1965.

40. Geel, S.E. and Timiras, P.S.: Influence of growth hormone on cerebral cortical RNA metabolism in immature hypothyroid rats. Brain Res., 22:63, 1970.

41. Krawiec, L., Garcia Argiz, C.A., Gomez, C.J. and Pasquini, J.M.: Hormonal regulation of brain development. III. Effects of triiodothyronine and growth hormone on the biochemical changes in the cerebral cortex and cerebellum of neonatally thyroidectomized rats. Brain Res., 15:209, 1969.

42. McNulty, W., Hagemenas, F., Deamer, N. and Kittinger, G.: The effect of chronic decapitation on the development of the rhesus fetus (Macaca mulata). 55th Annual Meeting of the Endocrine Society, Chicago, 1973.

43. Chang, C. and Scott, R.: Cerebral zinc content in the fetus (Macaca mulata). In: Cheek, D.B. (Ed) Fetal and Postnatal Growth in the Primate. New York, John Wiley & Sons, in preparation.

44. Holt, A.B., Mellits, E.D. and Cheek, D.B.: Comparisons between nucleic acids, protein, zinc, and manganese in rat liver: A relation between zinc and ribonucleic acid. Pediat. Res., 4:157, 1970.

45. Sandstead, H.H., Gillespie, D.D. and Brady, R.N.:
     Zinc deficiency: Effect on brain of the suckling
     rat. Pediat. Res., 6:119, 1972.

46. Giacobini, E.: A cytochemical study of the locali-
     zation of carbonic anhydrase in the nervous system.
     J. Neurochem., 9:169, 1962.

47. Korhonen, E. and Korhonen, L.K.: Histochemical
     demonstration of carbonic anhydrase activity in
     the eyes of rats and brain. Acta Ophthal.,
     (Kobenhavn), 43:475, 1965.

48. Schulte, F.J., Lasson, U., Parl, U., Nolte, R. and
     Jürgens, U.: Brain and behavioural maturation
     in newborn infants of diabetic mothers. II:
     Sleep cycles. Neuropädiatrie, 1:36, 1969.

49. Farquhar, J.W.: Prognosis for babies born to dia-
     betic mothers in Edinburgh. Arch. Dis. Childh.,
     44:36, 1969.

50. Frost, L.H. and Jackson, R.L.: Growth and develop-
     ment of infants receiving a proprietary preparation
     of evaporated milk with dextri-maltose and
     vitamin D. J. Pediat., 39:585, 1951.

51. Fomon, S.J.: A pediatrician looks at early nutrition.
     Bull. N.Y. Acad. Med., 47:569, 1971.

52. Mellits, E.D. and Cheek, D.B.: The assessment of
     body water and fatness from infancy to adulthood.
     Monog. Soc. Res. Child Develop., Serial No. 140,
     35:12, 1970.

53. Parra, A., Schultz, R.B., Graystone, J.E. and Cheek,
     D.B.: Correlative studies in obese children and
     adolescents concerning body composition and plasma
     insulin and growth hormone levels. Pediat. Res.,
     5:605, 1971.

54. Cheek, D.B., Schultz, R.B., Parra, A. and Reba, R.C.:
     Overgrowth of lean and adipose tissues in adoles-
     cent obesity. Pediat. Res., 4:268, 1970.

55. Stauffacher, W., Croffort, O.H., Jeanrenaud, B. and Renold, A.E.: Comparative studies of muscle and adipose tissue metabolism in lean and obese mice. Ann. N.Y. Acad. Sci., 131:528, 1965.

56. Felig, P., Marliss, E. and Cahill, G.F., Jr.: Plasma amino acid levels and insulin secretion in obesity. New Engl. J. Med., 281:811, 1969.

57. Forbes, G.B.: Lean body mass and fat in obese children. Pediatrics, 34:308, 1964.

58. Reba, R.C., Cheek, D.B. and Mellits, E.D.: Body composition studies: Growth of intracellular mass and metabolic size and the assessment of maturational age. In: James, A.E., Wagner, H.N. and Cooke, R.E. (Eds) Pediatric Nuclear Medicine. Philadelphia, W.B. Saunders Co., 1973.

59. Flynn, M.A., Woodruff, C., Clark, J. and Chase, G.: Total body potassium in normal children. Pediat. Res., 6:239, 1972.

60. Reba, R.C., Cheek, D.B. and Leitner, F.C.: Body potassium and lean body mass, p. 155. In: Cheek, D.B. (Ed) Human Growth. Philadelphia, Lea and Febiger, 1968.

61. Wiedemann, E. and Schwartz, E.: Suppression of growth hormone-dependent human serum sulfation factor by estrogen. J. Clin. Endocr., 34:51, 1972.

62. Migeon, C.J., Green, O.C. and Eckert, J.P.: Study of adrenocortical function in obesity. Metabolism, 12:718, 1963.

63. Knittle, J.L. and Hirsch, J.: Infantile nutrition as a determinant of adult adipose tissue metabolism and cellularity. Clin. Res., 15:323, 1967.

64. Hill, D.E., Hirsch, J. and Cheek, D.B.: The non-
     collagen protein in adipose tissue as an index of
     cell number. Proc. Soc. Exp. Biol. and Med.,
     140:782, 1972.

65. Salans, L.B. and Dougherty, J.W.: The effect of
     insulin upon glucose metabolism by adipose cells
     of different size. Influence of cell lipid and
     protein content, age, and nutritional state. J.
     Clin. Invest., 50:1399, 1971.

66. Cheek, D.B. and White, J.J.: Obesity in adolescents--
     the role of hormones and nutrition. IXth
     International Congress of Nutrition, Mexico City,
     1972.

67. Cheek, D.B.: Body composition, hormones, nutrition
     and adolescent growth. Conference on the Control
     of the Onset of Puberty, Airlie Foundation, Airlie,
     Virginia, 1972.

ACKNOWLEDGEMENT

    Thanks are expressed to S. Karger (Basel) for per-
mission to reproduce previously published material.

MATERNAL LEUKOCYTE METABOLISM IN FETAL MALNUTRITION

Jack Metcoff

From the Oklahoma University Health Sciences
Center, Oklahoma City, Oklahoma, USA

About 15 percent of successful pregnancies yield
low-birth-weight babies (less than 2500 g). It is
estimated that the low birth weight in about one-third
to one-half of these infants is not due to prematurity,
but is the result of fetal malnutrition (FM).[1] An in-
determinate number of surviving fetally malnourished
babies have permanent physical, neurologic, intelligence or learning defects. In addition, about one-
third of fetal deaths has been ascribed to fetal
malnutrition.[2] It has been suggested that maternal
malnutrition may represent a teratogenic insult during
the embryonic life of some infants.

Whether fetal malnutrition results from poor maternal nutrition or fetal, placental, other maternal
factors, or various combinations of these is not known.
We hypothesize that idiopathic fetal malnutrition is
essentially maternal in origin and may be related to
maternal nutrition. Thus, maternal cells, particularly
rapidly replicating cells, might illustrate phenomena
that also characterize replicating cells in the fetus.
According to this hypothesis, if the metabolism of
replicating fetal cells is impaired, similar metabolic
changes should be evident in the rapidly replicating
cells of the mother, particularly during the last 10 to
12 weeks of pregnancy in the human.

The prevalent view of fetal malnutrition holds
that cell replication is impaired in utero. Cell replication, of course, depends on protein synthesis and
is linked closely with energy metabolism. Protein
synthesis requires energy as well as a sufficient supply of nutrient substrates produced by essential enzymes. If significant changes in cell substrates or
enzyme dependent functions are characteristic of the
mother bearing an infant with fetal malnutrition, the
changes should serve as biochemical markers of fetal
malnutrition. If the cellular changes are a characteristic of the mother, not of the pregnancy, they
should persist after parturition and should be present
in some nonpregnant women.

Biochemical markers, indicating fetal malnutrition,
have not yet been reliably identified in pregnant
women. Various chemical substances in urine, amniotic
fluid, and the serum of pregnant women have been used
as indicators of fetal or placental status with regard
to successful outcome of the pregnancy. The tests for
these substances have been most useful with regard to
events that might cause fetal death, but have not contributed significantly to a prenatal diagnosis of fetal
malnutrition.

It should be emphasized that the term "fetal malnutrition" probably includes several different types of
processes that may occur in utero, depending on the
timing and nature of the adverse factors affecting the
mother before and/or during pregnancy. The effects of
these adverse factors on fetal growth and development
may lead to different types of fetal malnutrition (FM).
For many years, FM was thought to be the result of
"placental insufficiency"-- e.g., impaired protein or
metabolite synthesis and insufficient transport of
nutrients. This must be true in some instances where
the placenta is partially destroyed as a result of disease or interrupted blood supply. The placenta of FM
babies is small and has a reduced number of cotyledons[3]
and hypertrophic cells,[4] but the levels of energy metabolites, energy-related enzyme functions[5] and RNA and
protein synthesis[6,7] are very active and quantitatively
equivalent to those of the larger, normal baby placenta.

If fetal malnutrition represents an inadequate supply of nutrients to the developing organism, and cannot be ascribed solely to insufficient placental transport or other placental functions, what role does the mother's nutritional status play in fetal development? The evidence that maternal malnutrition inhibits or alters development of the human fetus to produce a malnourished infant is largely circumstantial.[8] In the experimental animal, it seems clear that fetal growth, as well as the development of specific organs, can be impaired by poor nutrition in the mother, especially by protein deprivation in the last half of gestation.[9]

NATURE OF THE PROBLEM

Definition of FM. Generally, babies born with manifestations of FM have been considered in England and on the Continent as "small for gestational age" or "small-for-dates." Of course, the estimate of gestational age may be inaccurate. The small-for-dates baby is underweight, and may be short as well, for whatever period of time it has grown in the uterus. Other common synonyms are "growth retarded in utero" or, in America, now frequently called "fetal malnutrition." The term "fetal malnutrition" was coined independently by Scott and Usher[10] and by Naeye.[11] Examining the tissues of these very small babies, Naeye drew an analogy between their tissue composition and that of children with severe protein-calorie malnutrition. Naeye concluded that organs of "small-for-dates" babies had the morphologic characteristics of malnutrition and that these changes must have developed in utero. In recent studies, Usher et al.[2] have shown that fetal malnutrition is an important cause of death among neonates. Naeye has suggested that maternal malnutrition is the most likely cause of fetal death when known complications of pregnancy are absent.

At the Babies' Hospital in New York, Naeye et al. performed 1,002 consecutive autopsy examinations on fetuses who were either stillborn or died in the perinatal period. Of these, 553 cases had known gestational periods with no abnormalities in the placenta or the

fetus or evident diseases in the mother to account for
fetal death. Eighty-three of the fetuses were markedly
underweight for their gestational age and showed morphologic changes in organ and cell structure that were
characteristic of undernutrition. Data on family incomes
revealed that all 83 of these babies were born to extremely poor families. The remaining infants for whom
family income was known (386) were not as small for gestational age, had significantly less alteration in organ
and cell structure, and additionally, were born to non-
poor families. These observations led Naeye et al. to
conclude that "maternal malnutrition during gestation
provides the simplest explanation for the undernutrition
found in newborn infants of the poor."[12]

This line of reasoning also led us to speculate
that maternal malnutrition was an important factor contributing to the high incidence of fetal malnutrition
noted by our research group among a low socioeconomic
class of pregnant women in Mexico.[13] Fetal malnutrition,
however, has not been related conclusively to maternal
nutrition either in Mexico or elsewhere. One of the important questions that must be answered is whether
indeed there is a relationship between human fetal malnutrition and the nutritional status of the mother.

Estimated Incidence

The incidence of fetal malnutrition is unknown.
The demographic data collected in the United States
indicate how many babies are born with birth weights less
than 2500 g and, hence, are considered low birth weight
babies; and how many babies are born too soon (before
38 weeks of gestation), i.e., prematurely. But individual births generally are not analyzed demographically
with regard to both birth weight and gestation age.
Thus, our current vital statistics do not indicate which
of the babies weighing 2500 g or less at birth have a
gestation period of 40 weeks, or 39 weeks, 38 weeks, etc.
The baby who is growing appropriately in utero for a full
gestation period of 40[+] one week should weigh more than
2500 g and should be between 46 and 52 cm long. If
weight, and sometimes length as well, are significantly

less than expected for any given gestational age, such a baby is considered small-for-dates, which usually is attributed to some form of fetal malnutrition.

Lacking specific data, I have attempted to estimate the magnitude of the problem of FM in our country (Table I). This estimate of the annual incidence of fetal malnutrition (FM) undoubtedly can be criticized, but I believe it provides an order of magnitude.

TABLE I

Estimated Annual Incidence of
Small for Gestational Age Births
(Fetal Malnutrition or F.M.)
in the United States, 1968[*]

ACTUAL:

Live births	3,501,564
Births under 2500g (LBW)	286,528
Fetal deaths	55,293

ESTIMATED:

% LBW $\cong$ F.M. = 33-50 (40)

LBW = F.M.	114,611

% Fetal deaths $\cong$ F.M. = 30

Fetal deaths = F.M.	16,588
Annual incidence F.M.	131,199

[*]Based on data in references 1, 2, 4, 11 and 14.

Eventually we should be able to get more specific figures. In 1968, there were about 3.5 million live births
in the USA. Of these, there were about 286,000 births
in which the baby weighed less than 2500 g at the time
of birth. There were also 55,000 fetal deaths.[14]
Gruenwald[1] and Scott and Usher[10] estimated that between
33 and 50 percent of low birth weight babies are not
"prematures" but have fetal malnutrition. In our own
small series, approximately 40 percent of the babies
born with weight less than 2500 g and below the 10th
percentile for expected gestational age at birth had
clinical characteristics of fetal malnutrition.[15] I
have chosen to use the average figure of 40 percent
which is placed in brackets. Using the 40 percent figure, 114,000 of the 286,000 low birth weight babies presumably would be infants with fetal malnutrition. Usher
concluded that 30 percent of the fetal deaths he reviewed
could be attributed to malnutrition of the fetus. If we
use that figure, then about 16,000 of the 55,000 fetal
deaths could be ascribed to fetal malnutrition. The
total annual incidence in the United States would be
131,000 babies per year either dying in utero from, or
born with, fetal malnutrition.

I suspect that the incidence of fetal malnutrition
may be proportionately higher in developing countries.
Thus we are dealing with an enormous public health problem that is not confined to developing or to underdeveloped countries but is also characteristic of highly
industrialized, technically advanced societies. Presumably, those fetuses with the most extreme forms of
malnutrition would be the most likely to die in utero,
whereas those with less extreme forms would be likely
to survive until birth.

SIGNIFICANCE OF FETAL MALNUTRITION

The consequences for the infant are uncertain.[16]
As illustrated in Table II, these may be divided into
two general categories: "Likely" and "Possible."

Among the <u>likely</u> consequences, the baby is small--
body and organs, including the brain. An interesting
feature, however, is that the brain, per unit of body
length or weight, is of average or increased size, although it may be small compared to the brain of a full
sized term baby of the same gestational age. The brain
may have a reduced number of cells (DNA content) but

TABLE II

Consequences of Intrauterine Malnutrition
for Infant

1. LIKELY
 A. Small Size
  1. Body
  2. Organs, including brain

 B. Body composition altered
  1. < Cell number (DNA)
  2. < Minerals
  3. < Nervous tissue lipids
  4. < Liver glycogen

 C. Congenital anomalies

2. POSSIBLE
 A. Impaired neonatal functions
  1. Neural
  2. Temperature regulation
  3. CHO metabolism (hypoglycemia)
  4. Kidney function
  5. Protein metabolism

 B. Postnatal effects
  1. Learning and behavior
  2. Neurointegration
  3. Growth

 C. Increased mortality
  1. Perinatal, neonatal, postneonatal

the number of cells per unit of brain mass seems to be
well preserved. However, DNA content reflects numbers
of both the supporting glial cells and the neuronal
cells in the brain. Recently, Chase[17] and Dobbing[18]
have shown that neuronal cell DNA in the cortex increases rapidly before birth and would be more likely
to be affected by prenatal malnutrition. Cerebellar
DNA increases before birth but the peak increase occurs
postnatally. Thus malnutrition during the first postnatal year would be likely to have greater impact on
the development of cerebellar neurones. Some recent
studies of the brain cortex from rats subjected to
fetal and weaning-period malnutrition, indicate reduction in the number of synaptic terminals of the
neurons.[19,20] The biochemical composition of these
synaptic terminals (synaptosomes) is affected also, having reduced DNA, RNA and phospholipids.[21,22] Also liver
glycogen may be reduced per unit of DNA at the time of
birth.[23] In human infants at least, this is thought to
be a major cause for the hypoglycemia that frequently
characterizes babies born with intrauterine malnutrition. The incidence of hypoglycemia is much greater in
these infants than in equally small infants born prematurely but appropriately grown for their shortened gestational period.[24] However, most small infants born
with fetal malnutrition are less likely to have a respiratory distress syndrome, which frequently afflicts
prematurely born babies with equivalent birth weight.
FM babies, born at term, tend to have appropriately
mature lungs. Infants born with fetal malnutrition
seem to have a higher incidence of congenital anomalies,
which, in some series, runs as high as 50 percent.[2]

     Possible consequences for the infant include alter-
ations at birth in neurological functions, temperature
regulation, carbohydrate metabolism and postnatal metabolism of protein. Postnatal effects may include
impaired capacity for learning[25,26] and behavioral responses and perhaps difficultues with neuro-integrative
performance.[17,25] The term "neurointegrative performance" refers to the capacity to perform integrated
tasks (that require synthesis of information) and kin-
esthetic performance in response to specific cues.[27]

Behavioral responses noted in the tables refer to social
behavior. In several series, there was little or no
"catch-up growth" among the FM babies, unlike "normal"
prematures or children who have suffered various types
of malnutrition during early life, who can show catch-
up growth. Although the term fetal malnutrition (FM) is
a useful label, there is no clearcut evidence to date
that the mothers of babies born with FM are malnourished.
One of the objectives of our current study is to find
whether or not a relationship exists between the nutritional status of the mother and the development of the
fetus.

Some of the factors that have been related to the
incidence of FM are presented in Table III. One of
these is the present and past nutritional status of the
mother.[8] Women who live and conceive at high altitudes
tend to have small babies. Apparently increased altitude decreases the size of the baby at birth. This
becomes important from an anthropometric point of view,
because a commonly used scale for estimating intrauterine growth in relation to gestational age was developed
in Denver[28] which has an altitude of about 6,000 feet.
These babies may have been smaller at birth than infants
born at lower altitudes and, hence, the Denver scale of
intrauterine growth may underestimate expected fetal
growth for infants at lower altitudes. Viral and bacterial infections and, as pointed out by Jelliffe, parasitic (malarial) infection of the placenta will lead to
FM. The timing of the infection during gestation may
be particularly important to the fetus and to the occurrence of congenital abnormalities.

Possible Public Health Significance

Bergner and Susser have reviewed the factors
associated with low birth weight infants.[8] In particular, they attempted to assess whether low birth weight,
including both prematurely born and small-for-dates (FM)
infants, could be attributed to events affecting that
specific pregnancy, or whether there were intrinsic
characteristics of the mother that affected all her

pregnancies. In the latter case, the same mother would
be likely to produce low birth weight babies in subsequent pregnancies. If some environmental, rather than
endogenous, factors influenced fetal development, it
would be easier to correct these and prevent FM. If
FM were the result of endogenous factors, it would be

TABLE   III

Factors Related to Incidence of
Intrauterine Malnutrition

I.   Maternal
          Socioeconomic status
          Race
          Disease
                 Toxemia
                 Hypertension
                 Bacteriuria
                 Viral infections
                 Drugs
          Nutritional status--present and past
          Smoking
          Pre-gravid weight and height
          Education

II. Fetal
          Length of gestation
          Birth rank
          Multiple birth
          Discordant twin
          Genetic factors

III. Fetomaternal Factors (Placental)
          Vascular--Blood flow
          Oxygenation--Altitude
          Hormones (P.L., GH, Ins)
          Placental transport (amino acids, minerals)

important to identify which pregnant women were likely to have FM babies in order to design and provide therapy for these women. In either case, prenatal diagnosis of FM is required.

Relation of FM to Maternal Nutrition

Some questions about nutritional factors that might cause fetal malnutrition are indicated in Table IV. Timing, with relation to the vulnerable period described by Dobbing,[29] may be particularly important. The term

TABLE IV

Questions RE Nutritional Causes of Intrauterine Malnutrition

A. Timing
    1. Relation to Critical Period

B. Environmental Influences
    1. Remote
        A. Grandmother's nutrition during
                pregnancy with mother
        B. Mother's childhood nutrition

    2. Current Pregnancy
        A. Before
        B. During
        C. Before + during
        D. Time and duration

C. Type and Extent of Deprivation
    1. Protein
    2. Calories
    3. Fats
    4. Combinations of 1-3
    5. Minerals
    6. Vitamins

"critical period" refers to the time when intrauterine
growth and replication of cells is occurring most rapidly; hence the cells may be more vulnerable. If
adverse influences occur at that time there may be a
greater chance for interference with cell replication.
Hence, events so timed would be more likely to limit
growth by limiting replication of cells.

Environmental influences may be remote or related
to the current pregnancy. "Grandmother's nutrition
during pregnancy with the mother" is a reference to
Zamenhof's experiment.[30] An F_0 generation of rats was
fed a restricted diet prior to mating and through 21 days
of gestation. Most of the rats born in the F_1 generation showed no evidence of fetal malnutrition. However,
when the F_1 females were mated and delivered their pups,
the F_2 generation, the pups were small and had some
characteristics of FM. Zamenhof concluded that if the
"grandmother" had suffered malnutrition prior to and
during her gestation, her "child" might escape being
fetally malnourished but the "grandchild" would not.
He also indicated that the mothers' childhood nutritional state was important because if the F_1 generation
is suckled by dams who were malnourished, their growth
would be impaired and they were more likely to produce
small babies when mated.

During the current pregnancy, the timing and severity of malnutrition is important. Of course, the
type of nutrient deficiency may be important also.
While we assume that there are cause and effect relationships between maternal nutrition and fetal development, there is no substantial proof of this in human
populations.

There are many different types of fetal growth
impairment.[15] Figure 1 illustrates three hypothetical,
but commonly observed, types.[31] These might be interpreted with respect to the timing, extent and degree of
intrauterine malnutrition. These curves are superimposed
on modifications of the Denver intrauterine growth chart.
All similar charts relate length and weight at birth to
gestational age. Of course, these curves were not

constructed by following the growth of a single baby.
Rather, many cross-sectional observations were made of
babies born at 28, 32, 36 weeks of gestation, and so on.
The percentile distributions of these measurements were
determined and a curvilinear relationship estimated between the points to construct these growth curves. The
actual logarithmic growth pattern over time, of the
individual fetus has not been established. These hypothetical growth curves indicate how timing of some
adverse event might affect fetal growth. For example,
in the lower curve, "Type I fetal malnutrition," the
baby is always below the 10th percentile in growth for
any gestational period. The baby starts small and
remains small. The discrepancy between expected weight

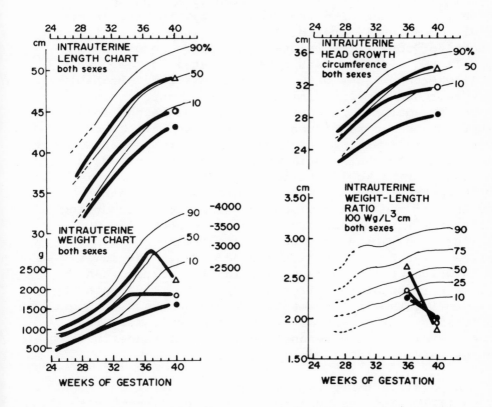

Figure 1. Hypothetical types of fetal growth retardation. Type I = ● ; Type II = ○ ; Type III = △ .

gain toward the end of gestation and actual weight gain
is very great.   This is the interpretation given to
observations by Ghosh et al.in Northern India with
respect to birth weights of a large series of babies
born in their hospital.[32]   At any birth weight, the
babies were always below the 10th percentile standards
for gestation age.   If the length, the weight and the
head circumference are proportionally small, the intrauterine growth pattern most likely would be like that
of a Type I baby.   This type of malnutrition probably
affects the fetus throughout most of the pregnancy.
Although Miller et al. have emphasized the use of the
ponderal index to identify fetal malnutrition,[33] usually
the ponderal (Rhorer) index of such a baby will be
within normal limits.

In Type II FM, there is a significant reduction of
weight, very much like the Type I baby, but the length
is at, or slightly greater than, the 10th percentile.
Length, of course, is related to skeletal growth.   In
Type II FM, linear growth appears sustained reasonably
well, whereas weight gain has been impaired.   Probably
this represents malnutrition of shorter duration.
There is a disproportionate reduction of weight for
length or head circumference.   For these babies, the
ponderal index will be reduced.   This sort of pattern
was thought to characterize the starvation period in
Holland, during World War II, when, for a period of
20-30 weeks, food deprivation approached starvation
levels.   If the last half of pregnancy occurred during
the starvation period, the babies were about 150-200 g
smaller than babies born before the famine or some two
months after the starvation period ended.   The lower
birth weight babies were only slightly shorter than
babies born before or after that period.   Thus, starvation during the last trimester of pregnancy was thought
to impair terminal intrauterine weight gain more than
linear growth of the fetus.   Head circumference in these
babies was not significantly reduced.

A possible third type of fetal malnutrition
(Type III) shows no impairment of linear growth nor
of head circumference, but the baby is born very much

underweight. The interpretation here is that some very
severe, acute event occurred toward the end of pregnancy
leading to severe malnutrition and utilization of subcutaneous fat with weight loss by the fetus. Bergner
and Susser suggest that in the seige of Leningrad the
severe starvation that occurred for about two months
would have affected babies in the last few weeks of
pregnancy most seriously.[8] Such babies would be underweight, but length would not be affected. Clinically,
these types of fetal malnutrition are seen as isolated
events, but the factors involved are completely unclear.
Prospective studies are lacking.

OBJECTIVES OF OUR PROJECT

Maternal Malnutrition and Fetal Development (MMFD)

Our MMFD study has five objectives:

A. Predict FM as early as possible in pregnancy.

B. Relate fetal growth and FM to the current
nutritional status of the mothers.

C. Attempt to discriminate between current environmental effects on mother/baby, and intrinsic characteristics of the mother (not related only to current
pregnancy) with respect to FM.

D. Attempt to prevent FM (e.g., enhance fetal
growth in utero) during the pregnancy of mothers identified as carrying an undergrown fetus.

E. Provide a study sample of infants with known
prenatal and birth status for future developmental
studies.

The first objective is to see if we can find some
way to predict whether a mother is carrying a baby with
fetal malnutrition. Are there any indicators in this
mother that would make it possible to determine that her

baby is going to be born with fetal malnutrition? Our
goal is to make this diagnosis as early as possible in
pregnancy. If we can achieve this, it might be possible to do something about it: e.g., to enhance fetal
growth in utero. If fetal malnutrition reflected poor
maternal nutrition and deficient nutrient intake during
pregnancy, then by feeding all mothers well and supple-
menting their diet during pregnancy perhaps we could do
away with fetal malnutrition. This, in fact, is one of
the objectives of the Guatemala intervention studies
reported by Habicht.[34] However, if there are 100,000,
among 3.5 million, pregnant women who will deliver
fetally malnourished babies in the United States, we
would have to supplement the diets of 3.5 million women
every year, indiscriminately. Even if it were practi-
cally possible to supplement the diet of all pregnant
women with calories, protein, minerals, etc., such
supplementation might be undesirable for many pregnancies in which the diet is either adequate or excessive.

STUDY DESIGN

The design of our study will make it possible to
obtain prospective information at specific times during
pregnancy. Our "Maternal Malnutrition and Fetal Development" (MMFD) study is supported by the National Institutes of Health (NIH), specifically, the National Institute of Child Health and Human Development. It involves
four groups of investigators and is carried out at four
institutions. The study is based at the University of
Oklahoma where some 30 individuals representing different disciplines are participating. Other components
are in Mexico City, at the Centro Medico Nacional of
the Instituto Mexicano del Seguro Social; the Human
Nutrition Research Laboratory of the U.S. Department of
Agriculture in Grand Forks, North Dakota under Dr. Har-
old Sandstead, and the Protein Research Laboratories of
the Department of Agriculture in Beltsville, Maryland
(Dr. Phillip McClain).

The study design is shown schematically in Figures 2A and 2B. Pregnant women before 16 weeks of

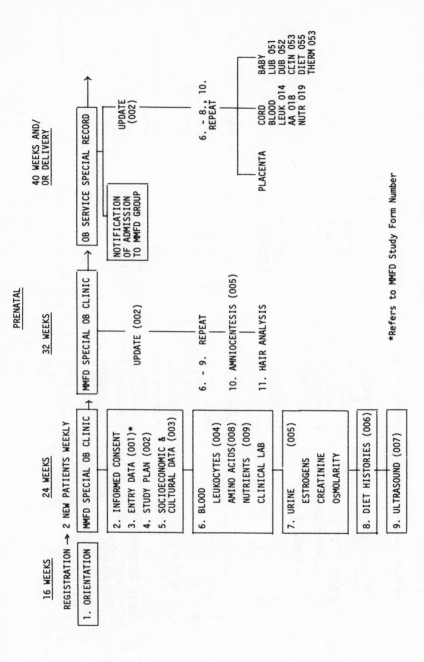

Figure 2A. Study design--maternal malnutrition and fetal development. Prenatal period.

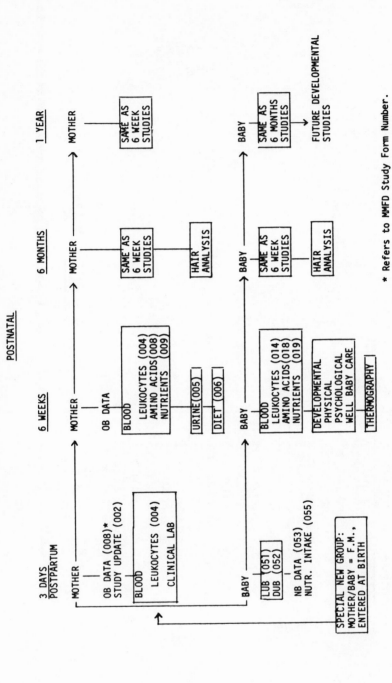

Figure 2B. Study design—maternal malnutrition and fetal development. Postnatal period.

gestation are registered for and are oriented to the
study. Less than 16 weeks of gestation is chosen as
the entry time because women are more likely to remember the date of their last menstrual period with reasonable accuracy and a base line estimate of uterine
fundal height can be used as a reference. At 24 weeks
of gestation, two of the previously registered patients
are entered into the study weekly in Oklahoma and currently three are entered into the study in Mexico.
Informed consent is obtained from these women at that
time. The nature and the details of the study to be
carried out with their approval, on them and their
babies, are explained. If they agree to participate,
the informed consent is signed and we proceed with the
study. The types of data collected are identified by
symbols indicating a computer form used for data entry.
The selection criteria in Oklahoma include women who
are free of known diseases, are having their first or
second pregnancies, and <u>therefore</u> are relatively young.
No attempt is made to select with respect to race or
socioeconomic group.

I would like to describe briefly some of the items
referred to in Figures 2A and 2B. The items on each
computer form comprise the data base for our study.
The entry data form (MMFD001) includes some 30 to 40
items relating to the demographic status of the women.
The study plan form (002) is a "yes"--"no" control
tabulation indicating the observations that have, or
have not, been made at the intended times. The third
form (MMFD003) details the socioeconomic and cultural
data. The 004 form describes the results of the leukocyte studies; the 008, the amino acids; the 009, the
blood and hair nutrient levels. A diet history is
obtained also by one nutritionist who uses a three day
diet record plus a 24 hour diet recall obtained at
each study point during pregnancy. Ultrasound fetal
encephalography measurements are reported on still
another form.

Arbitrarily, we have chosen to study the pregnant
patients at 24 weeks, 32 weeks and 40 weeks (delivery).

The study form is updated at each of these times. The
units six to nine, blood, urine, diet histories and
ultrasound measurements are repeated and an amniocen-
tesis is done if indicated for obstetrical reasons.
The information obtained has been useful in women who
have Rh incompatibilities or when fetal immaturity is
suspected. For hair analysis, we use scalp hair roots
for protein and DNA and the hair shafts are used for
analysis of trace minerals. At 40 weeks or delivery,
the patient is admitted to the Obstetrics Department
with a special record indicating that she is a study
patient, and our staff is notified. The 002 form is
updated, study units 6 to 10 are repeated on the mother
and then the products of pregnancy are entered into the
study. The placenta is examined, the cord blood is
obtained for leukocyte, amino acid, and nutrient
analyses.

The baby, of course, is assessed clinically and
some anthropometric measures are taken. The Lubchenco
evaluation of gestational age,[24] using some 27 criteria,
and the Dubowitz assessment of gestational age, which
uses 11 neurological and 11 clinical criteria,[35] are
carried out at 12 to 24 hours and again at three days
of age. Thus, for each baby, we have a gestational
history, ultrasound measurements of fetal head growth,
obstetrical measurements of the progress of pregnancy,
a clinical and anthropometric evaluation of the baby
and Lubchenco and Dubowitz estimates of gestational age.
Blood glucose is obtained at one hour of age from all
babies in the study. A specific diet is started at two
hours. Whole body thermography is done on the baby at
48 hours of age. The thermogram indicates heat emission
from the babies and is an index of subcutaneous fat,
including brown fat. Usually, thermograms are used in
clinical medicine for detection of cancer or for evidence of striking changes in peripheral circulation.
Our use of the thermogram is an adaptation. Standards
have not yet been established for evaluating thermograms in neonates but we plan to develop such standards
during our study. The thermogram will be related to
other measures of body composition, e.g., fatfold thickness, circumference measures, weight and length.

Additional observations are made three days post-
partum on the mother. The obstetrical data form and a
study update form are completed. Blood is obtained
again for leukocyte evaluations and clinical studies.
The baby is evaluated again by the Lubchenco and Dubowitz assessments. The newborn data are completed and
the record of the baby is then entered into the data
base. The nutrient intake to that time is recorded.

In Oklahoma, we anticipate about 10 babies will be
born with fetal malnutrition per year to the women
entered in our study. In Mexico, we anticipate about
twice that number of FM babies annually. In three
years, between 90 and 100 babies with fetal malnutrition should be born to prospectively studied mothers.
In addition, of course, there will be about twice that
number of normal premature babies. The majority of the
babies will be normal, full term babies. To supplement
our study with fetal malnutrition babies, we introduced
a "special study" group at the time of birth. This
group represents women who were not in the study on a
prospective basis but either by prenatal examination
appeared likely to have a small-for-dates baby, or at
delivery had a baby with fetal malnutrition. Cord
blood and the mother's peripheral blood are obtained,
baby and mother are entered into the study and thereafter are followed like other study patients. It should
be possible to enter an additional 30 FM special study
babies per year into our Oklahoma study. In Mexico,
two to three babies daily are born with clinical evidence of fetal malnutrition at the Centro Medico
Nacional. We hope to enter one or two of these babies
into the study per week which should give us an additional 50 FM Mexican babies per year.

At six weeks, all mothers and babies are again
studied in a special clinic. Blood, urine and diet
assessments are done on both mother and baby. Psychological and developmental testing on the baby is done
also. Thermography is repeated. Similar studies are
done at six months. The hair analyses from the mother
and baby are repeated. The studies are repeated again
at one year.

These postnatal follow-up studies should indicate
whether the changes found during pregnancy were characteristic of the mother or of the pregnancy. For example, should changes in the leukocyte enzymes or metabolism noted during the course of pregnancy revert back
to normal leukocyte patterns after the pregnancy, it is
likely that the leukocyte changes were associated with
the pregnancy. If, on the other hand, the changes persist postnatally, it would have to be assumed that the
leukocyte changes during pregnancy were characteristic
of the woman, and not of that particular pregnancy.
It is known that leukocyte metabolism may be modified
by the nutritional state of the person.[36] The dietary
pattern and nutrient status of each pregnant woman will
be quantified as noted previously. The Bayley scale
will be used to assess psychological/motor development
of the baby at six months and at one year of age.
Leukocyte and hair root nutrient studies also will be
done at these times.

The data base is established from these various
completed forms. All our material is promptly coded,
checked and turned over to the Data Manager, who re-
checks and submits the forms for key punching. Then
the data are entered into the computerized data base.
We use a system developed at the University of Oklahoma
called "GIPSY." This system does not require a predetermined format. GIPSY is an open-ended system which
allows one to ask the computer questions along any
format and it will return the information according to
the logic presented to it. The GIPSY system is based
upon a "label" dictionary. Currently, our dictionary
has a little over 1,000 labelled items. Our study requires that we develop mechanisms for easy access and
retrieval of the data. The data entry and retrieval
system lends itself to many different types of statistical analyses. Two epidemiologically oriented bio-
statisticians are participants in our study. They have
helped with study design and, of course, data analysis.
We use a "status report" system to get summary and correlation data each month.

LEUKOCYTES AS A MODEL CELL FOR STUDY

We are using the maternal leukocyte as a model cell:

1. To study pathways of leukocyte metabolism during pregnancy. Is there some characteristic leukocyte metabolic change during pregnancy?

2. To determine whether there is a characteristic metabolic pattern in the leukocyte during pregnancy, and whether deviations from this pattern occur? If so, could they serve as predictors of fetal malnutrition?

3. To determine the relation of maternal leukocyte metabolism to other features of pregnancy, e.g., toxemia. In Oklahoma City, toxemia of pregnancy occurs in about three quarters of the women who deliver babies with fetal malnutrition. In Mexico, three quarters of them do not have toxemia of pregnancy, but are more likely to have some underlying nutritional problem.

4. To study the relation of maternal leukocyte metabolism to the nutritional status of the mother.

5. To establish patterns of maternal leukocyte metabolism postnatally.

6. To relate the baby's leukocyte metabolism to that of the mother.

7. To relate the baby's leukocyte metabolism to its nutritional status at birth and postnatally.

8. To study the ontogenesis of the leukocyte metabolic pattern in the baby postnatally.

PRELIMINARY RESULTS

Some years ago, we observed that the levels of activity of two enzymes, adenylate kinase and pyruvate kinase, were reduced in leukocytes from the cord blood of babies born with fetal malnutrition. These

reductions of enzyme activities differentiated these
babies from others who were either born prematurely or
were normal full-term babies. Similar changes were
found in the leukocytes of their mothers at the time of
birth.[22] The first women who entered our current prospective study were sampled for the first time at
24 weeks of gestation in January, 1973. The average
age of the 42 women admitted to the program so far
(July 1973) is 22.4 ± 5.6 years. The mean nonpregnant
weight of these women is 59.3 kg. By repeating our
observations at 32 weeks and at 40 weeks, we will determine the weight gain during pregnancy. The mother's
height, as well as the height and weight of the father
are recorded. Most of the women in this preliminary
sample have had one pregnancy previously. A few items
illustrate the type of sociocultural-economic information we are obtaining (Table V). Only 13 of the 42
mothers smoked cigarettes and the smoking mothers averaged 14 cigarettes per day during pregnancy. Other
items include, for example, the level of education
attained by the mother and father in terms of elemen-
tary, high school and college years. The total family
income is recorded. We also determine languages spoken
at home, parentage, racial stock and so on.

     With respect to nutrition, our three-day food
record and 24-hour food recall indicate that of 19
women for whom the 24th week of gestation diet calculations have been completed, the average daily protein
intake was about 1 g/kg for nonpregnant weight.
Calorie intake averaged about 32 g/kg/day in these preliminary studies.

     Polymorphonuclear leukocytes are obtained from a
peripheral blood sample of the pregnant women. The
isolated leukocytes are analyzed for certain metabolic
characteristics. The leukocyte is used as a model for
other rapidly replicating cells, such as those in fetal
tissues. The polymorphonuclear leukocyte is the principal leukocyte in the peripheral blood of man. It
has a short life span which probably does not exceed
fourteen days.[37] During pregnancy, the polymorph
comprises about 70 to 75 percent of all the leukocytes

present in the peripheral blood. We obtain about 20 cc
of blood from the women at each of the study points
during and after pregnancy. Our isolation technique
yields about 95 percent polymorphonuclear leukocytes.
The final isolate contains $13-15 \times 10^6$ cells, having a
DNA concentration about 0.4-0.6 mg/ml homogenate.

The metabolism of the peripheral blood leukocyte
resembles that of other nucleated cells; however,

TABLE   V

Socio-Cultural Information at Entry (24 Weeks)

	N	Mean ± s.d.
Number of cigarettes smoked/day*	34	5.4 ± 9.1
Patient's education (years)		
Elementary	34	8.0
High school	33	3.2 ± 1.1
College	9	3.8 ± 3.2
Father's education (years)		
Elementary	34	7.9 ± 0.5
High School	30	3.7 ± 0.6
College	10	4.2 ± 3.7
Father's height (cm)	25	176 ± 8.2
Father's weight (kg)	28	72.8 ± 12.6
Total family income ($)	31	7,064 ± 5,891 (range:$750-$25,000)

*For the 13 mothers who smoked, the range was 1 to 30
cigarettes per day.

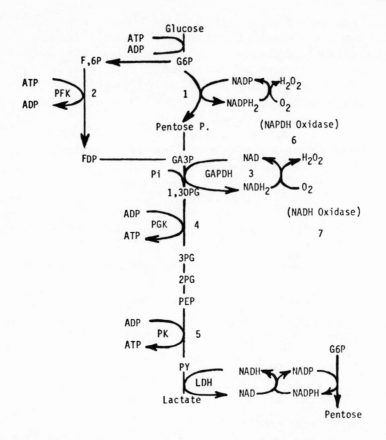

Figure 3. Pathways for glucose oxidation by the leukocyte.
Pentose phosphate shunt: . . . stimulated in phagocytosis
Phosphofructokinase (PFK):. . stimulated by ADP, AMP,
                               cyclic 3'5'AMP, P inhib-
                               ited by ATP, citrate
Glyceraldehyde phosphate:. . . stimulated by NAD, Pi
dehydrogenase (GAPDH):   inhibited by NADH, 1,
                               3DPG, cyclic AMP
Phosphoglycerokinase (PGK):. . stimulated by ADP,
                               inhibited by ATP
Pyruvic kinase (PK): . . . . . stimulated by G6P, FDP,
                               3PG, insulin
                               inhibited by ATP, NADH,
                               acetyl Co A
NAPDH oxidase
NADH oxidase   . . . . . stimulated in phagocytosis

glycolysis is the principal energy pathway (Figure 3).
The hexose monophosphate shunt, Krebs cycle, aerobic
mitochondrial oxidative metabolism, and protein synthesis also are present in the leukocyte (Table VI).
Peripheral blood leukocyte levels increase during pregnancy, reaching a plateau during the second trimester.[38]
Recently, Mitchell et al. reported that activities of
the hexose monophosphate shunt and of the enzyme myelo-
peroxidase in leukocytes are increased during human
pregnancy.[39] Usually these constitute the biochemical
drive for phagocytosis by the leukocyte; however,
phagocytic capacity was not increased during pregnancy.

TABLE   VI

Some Characteristics of the Human Blood Leukocyte

Short half life $\sim$ 14 days

Glycolysis predominates

Hexose monophosphate shunt

Krebs cycle

Aerobic mitochondrial oxidative metabolism

Protein synthesis

Reflects genetic defects, e.g.,
        Gangliosidosis
        Phosphorylase-deficient glycogen storage
        Fructose 6-diphosphatase
        Ataxia telangiectasia
        Chronic granulomatous disease

Reflects nutritional cell biochemical changes, e.g.,
        Protein-calorie malnutrition
        Vitamin C deficiency
        Lycine deficiency

Other studies have shown that specific enzyme defects or altered metabolite levels in the liver cells
of some genetically determined diseases, such as gan-
gliosidosis[40] and phosphorylase-deficient glycogen
storage disease,[41] are also present in the leukocyte.
Some amino acid deficiencies,[36] vitamin C deficiency,[42]
and altered protein synthesis[43] also are reflected by
the leukocyte. Leukocyte pyruvate kinase has kinetic
properties similar to that of the enzyme in muscle[44]
and the activity of the muscle enzyme is impaired by
protein-calorie malnutrition.[45] Levels of some glyco-
lytic and citric acid cycle metabolites and of adenine
nucleotides are altered significantly by protein-
calorie malnutrition.[46,47] Recently, leukocyte glycolysis and phagocytosis were noted to be impaired in
children suffering from protein-calorie malnutrition.[48]
These, and other observations, lend support to the thesis that metabolic changes in the leukocyte may be an
index of similar changes in the cells of other organs,[49] particularly if malnutrition occurs.

Recent work has suggested that the activities of
the enzymes adenylate kinase and pyruvate kinase normally increase in maternal leukocytes beginning around
the 34th week of pregnancy[7,50] (Figure 4). Because
the adenine nucleotides are present in the cytoplasm as
well as in mitochondria of cells, the rise of ATP and
ADP during the late trimester of pregnancy could reflect an increased content along with an increase in
cell size, rather than an increased concentration of
the nucleotides. Similarly, pyruvate kinase is a
cytoplasmic enzyme and although its activity is increased relative to nuclear mass (DNA), if cell size
increases in proportion, than the activities of the
enzyme per cell might be unchanged. Energy capacity
of the maternal leukocytes also increased during the
last trimester. As noted previously, energy capacity,
pyruvate kinase, and adenylate kinase activities are
reduced despite increased cell size in mothers deliv-
ering infants with fetal malnutrition.[13]

Some of the pooled regression data for the leukocytes, abstracted from our current study, are illustrated in Figure 5. These linear regression curves

were calculated from pooled regression coefficients for
cell size (protein/DNA ratio), adenylate kinase activity,
pyruvate kinase activity and the "energy charge" of the
leukocyte. The latter refers to the relationship between the adenine nucleotides as measured simultaneously in the leukocyte.   There seems to be an increase in cell size and in activity of these enzymes,
but a decrease in "energy charge" as gestation progresses. These changes suggest that we will find some
type of "normal" metabolic pattern in the leukocyte
that changes during pregnancy.

Recently we have determined some of the kinetic
properties of pyruvate kinase in leukocytes of pregnant women and of mothers at parturition. A preliminary assessment of the Michaelis-Menten constant (Km)
derived from Lineweaver-Burk plots of the enzyme-
substrate complex changes is shown in Figure 6. The
points for each curve indicate the mean values for
initial velocities found at each substrate concentration

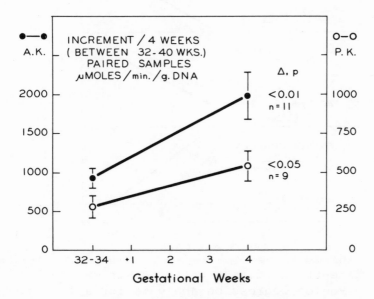

Figure 4.   Adenylic and pyruvic kinase in maternal
leukocytes in pregnancy.

in each group of women: mothers who delivered fetally
malnourished infants (FM = 6), mothers who delivered
normal full term infants (N = 6), and prenatal samples
from pregnant women between 29 and 38 weeks of gestation (PN = 10). The regression lines were calculated.
Reported values[51, 52] for the Michaelis-Menten constant
(Km) for pyruvate kinase range from 0.35 to 0.06 M x 10^{-3}.
In human leukocytes PK_{Km}PEP and ADP were found to be
0.1 and 0.18 x 10^{-3}M.[52] These values are closely

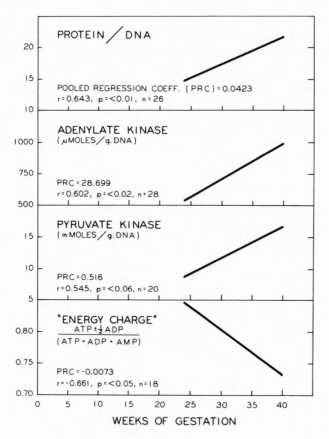

Figure 5. Time-trends during gestation in maternal
leukocytes. Pooled regression analysis for all subjects
with at least two studies between 23 and 41 weeks of
gestation.

approximated by the values we observed in leukocytes from women during pregnancy or at term--Km_{PEP} = 0.35 - 0.83 $\times$ $10^{-3}M$, Km_{ADP} = 0.12 - 0.49 X $10^{-3}M$.   At term

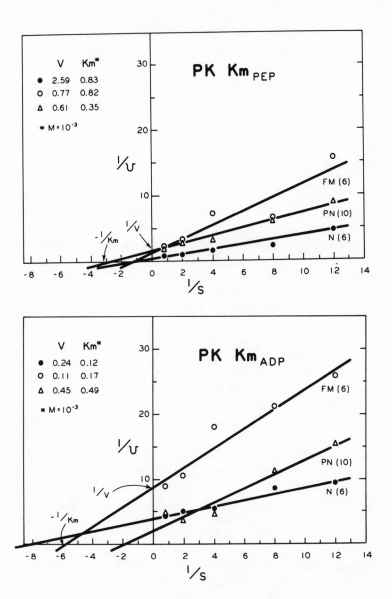

Figure 6.   PK Km_{PEP} and PK Km_{ADP} values for maternal leukocytes.

the apparent Km_{PEP} is higher; and for ADP lower. There
is a distinct difference in the slopes of the curves
for the FM mothers in comparison with normal mothers
for both substrates. The difference is attributable
largely to the lower V_{max} value for the enzyme in FM
mothers. This suggests either the presence of some
inhibitor or a reduction of enzyme protein and/or binding sites. The difference in apparent Km's also might
be attributable to some type of inhibitor. These preliminary data do not warrant firm conclusions, particularly considering that purified enzyme was not used
for these kinetic studies. Nonetheless, the differences between the groups, particularly that of the FM
mothers, extends our previous observations of pyruvic
kinase differences in the FM group.

RNA synthesis is dependent on the activity of DNA-
dependent RNA polymerase. DNA polymerase activity, in
turn, is depressed by many factors, including starvation.[53] Activating factors include K^+, cyclic AMP,
ribosome fractions, and acidic protein extracts containing protein kinase. Protein synthesis is ini-
tiated when ribosomal sub-units in the cytoplasm form
complexes with messenger RNA, initiation factors, and
transfer-linked amino acids. RNA polymerase probably
is not a single enzyme, but operationally may be considered as a complex of polymerases that are Mg^{++}
dependent, and Mn^{++} dependent and Mn^{++} dependent but
inhibited by α amanitin. Different controlling factors and inhibitors affect each of the polymerase
enzymes. Variables affecting the rate of RNA synthesis
may operate at any of the steps, including binding,
initiating, elongating, and terminating the polypeptide
chain. Important variables include nucleotide tri-
phosphate concentration, ionic strength, Ca^{++}, Mg^+, and
Zn^+ concentrations, etc.[54] The activity of DNA-
dependent RNA polymerase (essential for protein synthesis) was found to increase progressively in maternal
leukocytes from about the 28th week of gestation to
term. However, the mothers who delivered low-birth
weight babies had lower levels of RNA polymerase activities in their leukocytes during this same period[7]
although cell size was increased.[7,13,50] This

reinforces the thesis that the metabolism of the maternal leukocyte might reflect fetal development.   RNA
polymerase activity of leukocytes from the peripheral
blood of the mothers appeared to be correlated with the
birth weights of their infants[7] (Figure 7).   Activity
of the enzyme in leukocytes increased during gestation,
but mothers who eventually delivered low birth weight
infants (premature or small-for-dates) had low levels.

DNA dependent-DNA polymerase activity, considered
essential for DNA synthesis, in maternal leukocytes
seemed to be relatively constant between 28 to 40 weeks
of gestation.   There was only one exception noted at
38 weeks among these 13 mothers.   That mother had a
very high level of activity.   She delivered a fetally
malnourished baby two weeks later[30] (Figure 8).   This
isolated preliminary observation is particularly

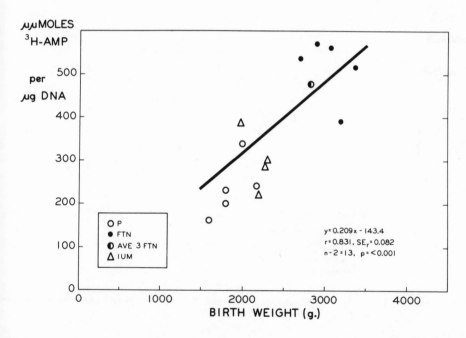

Figure 7.   RNA polymerase in maternal leukocytes related
to the birth weight of the infant.

interesting in view of the demonstration by Brasel
that maternal protein restriction during pregnancy
inhibits placental DNA synthesis in the rat, and this
inhibition is preceded by a reduction in DNA polymerase
activity.[55]

Finally, to illustrate still another kind of
approach that we are using, Figure 9 represents a preliminary asymetrical correlation matrix in which some
of our current data have been incorporated. The items
on the abscissa may be compared with those noted on
the ordinate. A number in a square represents the
number of paired observations for which a statistically
significant ($p < 0.05$) correlation was found. For
example, pyruvic kinase activity in the leukocytes
correlates with the size of the leukocytes. So does
adenylate kinase activity, and the adenylate kinase
activity also is correlated with the pyruvic kinase
activity, etc. There appears to be a statistically

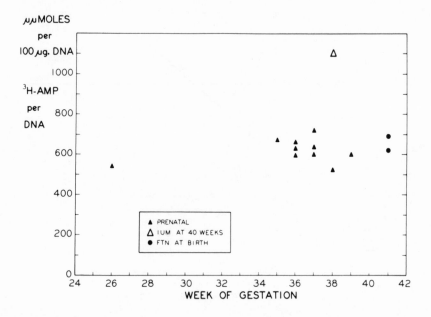

Figure 8.   DNA polymerase in maternal leukocytes.

significant positive correlation between diet calories and diet protein. So far, we also seem to have a correlation between the diet protein intake of the mother and fetal head size as determined by ultrasound encephalography at 24 weeks of gestation. Many such

ITEM	PROTEIN/DNA	PK	AK	PFK	G-6-PDH	Ecap	RNAp I(r)	RNAp II(m)	DIET PROTEIN	DIET CALORIES
PROTEIN/DNA										
PK	36[†]									
AK	39	37								
PFK										
G-6-PDH		37		39						
Ecap			35	34						
RNAp I(r)										
RNAp II(m)							9			
DIET PROTEIN										
DIET CALORIES								-4	14	

Figure 9. Correlations in maternal leukocytes at 24 weeks of gestation. Asymmetrical correlation matrix with missing data--to 7/12/73 MMFD (Okla.). The figures show the number of pairs correlated among 43 patients, where p   0.05. All the correlations are positive except as noted.

correlations either will evolve or will change as we
get more data. We do not want to interpret these results at this very early stage of our study. They are
presented to show various ways in which we plan to use
our data base and to indicate that many types of biostatistical techniques can be applied to assess the
relation between nutritional status of the mother and
the outcome of the pregnancy.

The term "nutritional status" is often used without a clear statement of its meaning. I would like to
digress somewhat from this discussion to attempt a
definition that is applicable not only to our study,
but might encompass the work presented at this conference.

"Nutritional Status" is an <u>operational term</u> that
implies measurements of the response to nutrients, or
lack of nutrients, by an individual, a particular
category of individuals, or a community. The precise
definition of nutritional status <u>changes</u> with the purpose of the evaluation and the modalities and responses
being measured; e.g., incidence of severity of acute
malnutrition as a result of starvation (siege of Leningrad, Biafra, etc.) versus subtle evidence of malnutrition (Fe deficiency) in a general population (Ten-
State Nutrition Survey in the USA). Therefore, a
single, all encompassing definition is likely to be so
general, it will not be helpful. The use of the term
at this conference has both <u>static (e.g., body size)</u>
and dynamic(e.g., work capacity, O_2 consumption,
<u>nutrient loading tests) connotations</u> regarding identi-
fication of changes that had occurred or were likely to
occur. The choice of either <u>"external" body measure-</u>
<u>ments</u> or <u>"internal" compositional measurements</u> must be
made in the <u>context</u> of the problem. The same given
value for some measurements could mean either nutritional deterioration or improvement, depending on
whether the observation was made on the downslope or
upturn of the curve of nutritional status. For example,
a markedly reduced arm circumference measurement or
plasma protein level could be a definite improvement
over a previous measurement--indicating the person was

improving; or it could be less, indicating that the malnutrition was getting worse. Similarly the same value in different populations may reflect genetic influences as well as environmental factors (e.g., birth weights in different populations). The data could refer to a single point in time on the one hand, or indicate changes with elapsed time, or with altered conditions, on the other hand. In some instances, the measurements relate observed values to some other variable (leukocyte measurement during pregnancy); in other cases, insights can be provided about the mechanisms through which the changes being measured had occurred (DNA content of the placenta, goiter, etc.).

Many public health problems have a nutritional component. Items relevant to public health also may include static demographic observations like mortality, or birth weights, and dynamic changes like physical and/or neurointegrative development and performance. To date, clear-cut relations between public health problems and nutritional status have been defined only under rather extreme conditions. It is suspected that nutritional status, superimposed on the genetic features of an individual, or category of individuals, largely determines growth and performance patterns and modifies responses to ecologic challenges. The relations between growth, health, response to environmental factors and nutritional status are being identified. The mechanisms by which nutritional state modifies the growth patterns and organ responses are beginning to be studied. In each instance, an array of measurements is usually required to describe the impact of nutritional changes. It should be emphasized that the selection of measures will depend on the purpose of the evaluation, but generally should involve some mix of "external" body and/or compositional measurements, and some "internal" organ system functional measurements repeated over a given time span. Ultimately, the functional measurements must be extended to include reactions at the cellular and the molecular level.

In the context of our studies, the purposes of determining nutritional status are to relate fetal

growth to maternal nutrition, and to make a prenatal
diagnosis of FM using the metabolism of the maternal
leukocyte as a cellular or molecular "marker." Of
course, if an antenatal diagnosis of FM can be made
early enough during pregnancy and correlated with
nutrient deficiency in the mother, it should be a relatively easy matter to provide the supplementation required for cell replication and growth of the fetus.
In this way, the unfortunate consequences of FM impair-
ing postnatal growth and performance might be prevented. The preliminary data derived from our prospective study offer some hope that our objectives may be
attained.

SUMMARY

Small-for-dates babies generally show evidence of
fetal malnutrition (FM). FM babies are likely to have
poor physical and intellectual development postnatally,
and a high incidence of congenital anomalies. While
the incidence of FM is unknown, it is estimated to
compromise 3 to 5 percent of all pregnancies in the
USA. Whether poor maternal nutrition is a major factor
in human FM is unknown. Prevention of FM will be more
effective than postnatal treatment, since replication
of cells in some organs, like brain cortex, is about
two thirds completed before birth. Prenatal diagnosis
is required to effectively prevent FM. This paper
describes an interdisciplinary, prospective study of
pregnant women, designed to determine the relation
between maternal nutrition and the outcome of pregnancy.
The metabolism of the maternal peripheral blood leukocyte serves as a model for rapidly replicating fetal
cells. It is anticipated that deviation from normal
trends of leukocyte metabolism during pregnancy will
serve as predictors of FM. Preliminary observations
are given that support the hypothesis that some metabolic features in the maternal leukocyte may serve as
"markers" for fetal growth. Some preliminary outputs
from our computerized data base also suggest that
correlations will evolve between determinants of the
nutritional status of the mother, fetal growth and

leukocyte metabolism. If the studies are successful, it should be possible to determine FM in utero and improve fetal nutrition to facilitate growth and cell development.

REFERENCES

1. Gruenwald, P.: Infants of low birth weight among 5,000 deliveries. Pediatrics, 34:157, 1964.

2. Usher, R.H.: Clinical and therapeutic aspects of fetal malnutrition. Ped. Clin. N. Amer., 17: 169, 1970.

3. Laga, E.M., Driscoll, S.G. and Munro, H.N.: Comparison of placentas from two socioeconomic groups. I. Morphometry. Pediatrics, 50:24, 1972.

4. Winick, M.: Cellular growth of human placenta. III. Intrauterine growth failure. J. Pediat., 71:390, 1967.

5. Rosado, A., Bernal, A., Sosa, A., Morales, M., Urrusti, J., Yoshida, P., Frenk, S., Velasco, L., Yoshida, T. and Metcoff, J.: Human fetal growth retardation. III. Protein, DNA, RNA, adenine nucleotides and activities of the enzymes pyruvic and adenylate kinase in placenta. Pediatrics, 50:568, 1972.

6. Laga, E.M., Driscoll, S.G. and Munro, H.N.: Comparison of placentas from two socioeconomic groups. II. Biochemical characteristics. Pediatrics, 50:33, 1972.

7. Metcoff, J., Wikman-Coffelt, J., Yoshida, T., Bernal, A., Rosado, A., Yoshida, P., Urrusti, J., Frenk, S., Madrazo, R., Velasco, L. and Morales, M.S.: Energy metabolism and protein synthesis in human leukocytes during pregnancy and in placenta related to fetal growth. Pediatrics, 51:866, 1973.

8. Bergner, L. and Susser, M.W.: Low birth weight and prenatal nutrition: An interpretative review. Pediatrics, 46:946, 1970.

9. Zamenhof, S., Van Marthenus, E. and Grauel, L.: DNA (cell number) and protein in neonatal rat brain: alteration by timing of maternal dietary protein restriction. J. Nutr., 101:1265, 1971.

10. Scott, K.E. and Usher, R.: Fetal malnutrition: its incidence, causes and effects. Amer. J. Obstet. Gynec., 94:951, 1966.

11. Naeye, R.L.: Malnutrition: a probable cause of fetal growth retardation. Arch. Path., 79:284, 1965.

12. Naeye, R.L., Diener, M.M., Harcke, H.T., Jr. and Blanc, W.A.: Relation of poverty and race to birth weight and organ and cell structure in the newborn. Pediat. Res., 5:17, 1971.

13. Metcoff, J., Yoshida, T., Morales, M., Rosado, A., Urrusti, J., Sosa, A., Yoshida, P., Frenk, S., Velasco, L., Ward, A. and Al-Ubaidi, Y.: Biomolecular studies of fetal malnutrition in maternal leukocytes. Pediatrics, 47:180, 1971.

14. Vital Statistics of the United States, 1968. U.S. Department of Health, Education and Welfare, Public Health Service, Washington, 1970.

15. Urrusti, J., Yoshida, P., Velasco, L., Frenk, S., Rosado, A., Sosa, A., Morales, M., Yoshida, T. and Metcoff, J.: Human fetal growth retardation. I. Clinical features of sample with intrauterine growth retardation. Pediatrics, 50:547, 1972.

16. Drillien, C.M.: The small-for-date infant: etiology and prognosis. Ped. Clin. N. Amer., 17:9, 1970.

17. Chase, H.P.: The effects of intrauterine and
     postnatal undernutrition on normal brain de-
     velopment. Ann. N.Y. Acad. Sci., 205:231, 1973.

18. Dobbing, J.: The later development of the central
     nervous system and its vulnerability. In:
     Davis, J.A. and Dobbing, J. (Eds) Scientific
     Foundations of Paediatrics. London, Wm. Heine-
     mann Medical Books, 1973.

19. Cragg, B.G.: The development of cortical synapses
     during starvation in the rat. Brain, 95:143,
     1972.

20. Bass, N.H., Netsky, M.G. and Young, E.: Effect
     of neonatal malnutrition on developing cerebrum.
     Arch. Neurol., 23:289, 1970.

21. Gambetti, P., Autilio-Gambetti, L., Gonatas, N.K.,
     Shafer, B. and Stieber, A.: Synapses and malnu-
     trition. Morphological and biochemical study
     of synaptosomal fractions from rat cerebral
     cortex. Brain Res., 47:477, 1972.

22. Bernal, A., Morales, M., Chew, S. and Rosado, A.:
     Effect of intrauterine growth retardation on
     the biochemical maturation of brain synaptosomes
     in the rat. Personal communication.

23. Shelley, H. J.: Glycogen reserves and their
     changes at birth and in anoxia. Brit. Med.
     Bull., 17:137, 1961.

24. Lubchenco, L.O. and Bard, H.: Incidence of hypo-
     glycemia in newborn infants classified by birth
     weight and gestational age. Pediatrics, 47:
     831, 1971.

25. Eaves, L.C., Nuttall, J.C., Klonoff, H. and Dunn,
     H. G.: Developmental and psychological test
     scores in children of low birth weight.
     Pediatrics, 45:9, 1970.

26. Fitzhardinge, P.M. and Steven, E.M.: The small-
     for-date infant. II. Neurological and intel-
     lectual sequelae. Pediatrics, 50:50, 1972.

27. Cravioto, J., DeLicardie, E.R. and Birch, H.G.:
     Nutrition, growth and neurointegrative devel-
     opment: an experimental and ecologic study.
     Pediatrics, 38:319, 1966.

28. Lubchenco, L.O., Hansman, C., Dressler, M. and
     Boyd, E.: Intrauterine growth as estimated
     from liveborn birth-weight data at 24 to 42
     weeks of gestation. Pediatrics, 32:793, 1963.

29. Dobbing, J.: Vulnerable periods in developing
     brain. In: Davison, A.N. and Dobbing, J.G.
     (Eds) Applied Neurochemistry. Oxford, Blackwell
     Sci. Publ., Ltd., 1968.

30. Zamenhof, S., Van Marthenus, E. and Grauel, L.:
     DNA (cell number) in neonatal brain: second
     generation (F_2) alteration by maternal (F_0)
     dietary protein restriction. Science, 172:
     850, 1971.

31. Metcoff, J.: Biochemical marker of intrauterine
     malnutrition. In: Winick, M. (Ed) Nutrition
     and Fetal Development. New York, John Wiley &
     Sons (in press).

32. Ghosh, S., Bhargava, S.K., Madhavan, S., Taskar,
     A.D., Bhargava, V. and Nigam, S.K.: Intra-
     uterine growth of North Indian babies.
     Pediatrics, 47:826, 1971.

33. Miller, H.C. and Hassanein, K.: Diagnosis of
     impaired fetal growth in newborn infants.
     Pediatrics, 48:511, 1971.

34. Habicht, J.-P.,Yarbrough, C., Lechtig, A. and
     Klein, R.E.: Relation of maternal supplemen-
     tary feeding during pregnancy to birth weight
     and other socio-biological factors. In:

Winick, M. (Ed) Nutrition and Fetal Development. New York, John Wiley & Sons (in press).

35. Dubowitz, L.M.S., Dubowitz, V. and Goldberg, C.: Clinical assessment of gestational age in the newborn infant. J. Pediat., 77:1, 1970.

36. Leise, E.M., Morita, T.N., Gray, I., LeSane, F. and Rodriquez, M.: Leukocyte enzymes as indicators of nutritional deficiency. Biochem. Med., 4:347, 1970.

37. Galbraith, P.R., Chikkappa, G. and Abu-Zahra, H.T.: Patterns of granulocyte kinetics in acute myelogenous and myelomonocytic leukemia. Blood, 36: 371, 1970.

38. Mitchell, G.W., Jr., McRipley, R.J., Selvaraj, R.J. and Sbarra, A.J.: The role of the phagocyte in host-parasite interactions. IV. The phagocytic activity of leukocytes in pregnancy and its relationship to urinary tract infections. Amer. J. Obstet. Gynec., 96:687, 1966.

39. Mitchell, G.W., Jr., Jacobs, A.A., Haddad, V., Paul, B.B., Strauss, R.R. and Sbarra, A.J.: The role of the phagocyte in host-parasite interactions. XXV. Metabolic and bacteriocidal activities of leukocytes from pregnant women. Amer. J. Obstet. Gynec., 108:805, 1970.

40. Holmes, B., Page, A.R. and Good, R.A.: Studies of the metabolic activity of leukocytes from patients with a genetic abnormality of phagocytic function. J. Clin. Invest., 46:1422, 1967.

41. Williams, H.E. and Field, J.B.: Low leukocyte phosphorylase in hepatic phosphorylase-deficient glycogen storage disease. J. Clin. Invest., 40:1841, 1961.

42. Loh, H.S. and Wilson, C.W.: Relationship between
     leucocyte and plasma ascorbic acid concentra-
     tions. Brit. Med. J., iii:733, 1971.

43. Gordon, R.O., Oppenheim, J.J., Souther, S.G. and
     Spinson, E.B.: Immediate in vitro leucocyte
     DNA synthesis: an early indicator of heart
     allograft rejection. Surg. Forum, 22:258, 1971.

44. Koler, R.D. and Vanbellinghen, P.: The mechanism
     of precursor modulation of human pyruvate
     kinase I by fructose diphosphate. In: Weber,
     G. (Ed) Advances in Enzyme Regulation 6:127,
     Oxford, Pergamon Press, Ltd., 1968.

45. Metcoff, J., Frenk, S., Yoshida, T., Torres-
     Pinedo, R., Kaiser, E. and Hansen, J.D.L.:
     Cell composition and metabolism in kwashiorkor
     (severe protein-calorie malnutrition in chil-
     dren). Medicine, 45:365, 1966.

46. Yoshida, T., Metcoff, J., Frenk, S. and de la
     Pena, C.: Intermediary metabolism and adenine
     nucleotides in leukocytes of children with
     protein-calorie malnutrition. Nature, 214:
     525, 1967.

47. Yoshida, T., Metcoff, J and Frenk, S.: Reduced
     pyruvic kinase activity, altered growth pat-
     terns of ATP in leukocytes, and protein-calorie
     malnutrition. Amer. J. Clin. Nutr., 21:162,
     1968.

48. Selvaraj, R.J. and Bhat, K.S.: Metabolic and
     bactericidal activities of leukocytes in
     protein-calorie malnutrition. Amer. J. Clin.
     Nutr., 25:166, 1972.

49. Jemelin, M. and Frei, J.: Leukocyte energy metabo-
     lism. III. Anaerobic and aerobic ATP produc-
     tion and related enzymes. Enzym. Biol. Clin.,
     11:289, 1970.

50. Yoshida, T., Metcoff, J., Morales, M., Rosado, A.,
     Sosa, A., Yoshida, P., Urrusti, J., Frenk, S.
     and Velasco, L.: Human fetal growth retardation.
     II. Energy metabolism in leukocytes. Pediatrics,
     50:559, 1972.

51. Taylor, C.B., Morris, H.P. and Weber, G.: A com-
     parison of the properties of pyruvate kinase
     from hepatoma 3924-A, normal liver and muscle.
     Life Sciences, 8 (Part II):635, 1969.

52. Campos, J.O., Koler, R.D. and Bigley, R.H.:
     Kinetic differences between human red cell and
     leukocyte pyruvate kinase. Nature, 208:194,
     1965.

53. Onishi, T.: Studies on the mechanism of decrease
     in the RNA content in liver cells of fasted rats.
     II. The mechanism of starvation-induced decrease
     in RNA polymerase activity in the liver.
     Biochem. Biophys. Acta, 217:384, 1970.

54. Burgess, R.R.: RNA polymerase. Ann. Rev. Biochem.,
     40:711, 1971.

55. Brasel, J.A.: Newer tools for the diagnosis of
     malnutrition, Ped. Annals, 2:18, 1973.

ACKNOWLEDGMENT

     The biochemical data reported here result from the
excellent technical efforts of G. Burns, S. Beck, S.
Cook, K. Fowler, J. Page, R. Theimer and I. Smith.
The laboratory was supervised by M. Mameesh, Ph.D. The
leukocyte protein and RNA synthesis measurements were
performed by G. Jacobson, Ph.D. W. Crosby, M.D. has
been in charge of obstetrical aspects and successful
patient coordination was due to K. Murphy. The diet
assessments were made by S. Perkins and B. Bradshaw.
K. Corff, R. N., performed the baby measurements and
assessments. W. Reid designed the computer format,
while A. Wilson developed many of the computer programs.

The biostatistical expertise that contributed to the
study was provided by P. Costiloe, M.S. and P. Anderson, Ph.D. In addition, M. A. Munter and D. Pierce
gave important administrative and secretarial help.
Progress was dependent on the devoted efforts of this
large team.

INTERRELATIONSHIPS BETWEEN BODY SIZE, BODY COMPOSITION AND FUNCTION*

Jana Pařízková

From VÚ FTVS, Charles University, Prague, Czechoslovakia

Growth acceleration and increase in body size are among the most spectacular changes that have occurred in human physique. These secular changes are considered to be due to improved nutrition and substantial limitation of infections during early periods of life in developed countries. Most morphological characteristics are now achieving greater values. During the last two decades, data from different countries indicate a greater increase in recumbent length and height than in measures of weight or breadth.[1,2] More detailed data on other morphological changes are lacking. For example, it is not known whether body components such as lean body mass and depot fat are increasing proportionally or whether other organs increase their size at the same rate as total body mass which applies for heart muscle.

Results from studies of work physiology suggest that the acceleration of morphological development is not parallelled by corresponding development of functional efficiency.[3] An increase in the number of diseases of

* The author thanks Dr. Wayne D. van Huss for help in the statistical analysis of data and use of the computer services at Michigan State University, East Lansing, Michigan, USA.

civilization, mostly of the cardiovascular system that originate in the pediatric period,[4] is a further reason why the various secular changes in human development, due to changes in ecological conditions in advanced countries, are not always favorable.

THE IMPACT OF NUTRITIONAL FACTORS

DURING EARLY ONTOGENY

As shown by McCance and Widdowson,[5] the influence of nutrition is not identical in various periods of life. Knittle and Hirsch[6] showed that very early nutritional manipulations can lead to permanent changes in the number of fat cells. Consequently, differences in morphology in adults are attributable partly to interventions during early ontogeny.[7] Later effects of various stimuli, including nutrition, could be partly responsible for other changes, e.g., the development of obesity or leanness in individuals with similar or identical conditions of life.[8,9] Some of our previous studies suggest that such interventions are possible even sooner, i.e., during last weeks of pregnancy.

In fullterm newborns from normal healthy mothers, measured during 4 to 48 hours after birth, the fatfold thickness (Figure 1) varied to a similar extent as in much older children (C.V. = 10 percent or more). Sexual differences were already apparent with higher values for suprailiac fatfold thicknesses in girls.[10] In children born before term (7th month), subcutaneous fat was nearly absent. In newborns of diabetic metabolically decompensated mothers with hyperglycemia not only body weight was increased, but also subcutaneous fat (Figure 1). Hoet[11] found a significant correlation between the body weight of newborns of diabetic mothers and the glucose level in umbilical venous blood. Measurements of Přibylová and Znamenáček[12] showed that the newborns of properly treated diabetic mothers without hyperglycemia had only slightly increased body weight, which indicates that the amount of fat was probably close to normal. These findings concerning the inborn variability of subcutaneous fat in normal children, and its increase in newborns of hyperglycemic

diabetic mothers, suggest that prenatal factors influence
the development of body fat.[7] Varying levels of blood
glucose during the last weeks of pregnancy, even when not
pathological, e.g., due to relative or absolute overfeeding by the mother, could be of decisive importance.

The influence of nutrition on the fetus is apparent
in animals in which different numbers of offspring are
born in a litter. For example, newborn rats vary significantly in birth weight depending on the number in a
litter. As a rule, the birth weight decreases as the
litter size increases.[5,13] This influence can remain
manifest for a long time. Our longterm experiments
usually showed a marked variability in body weight and
percent fat even if, from immediately after birth, male
rats were suckled in the nests in litters of uniform

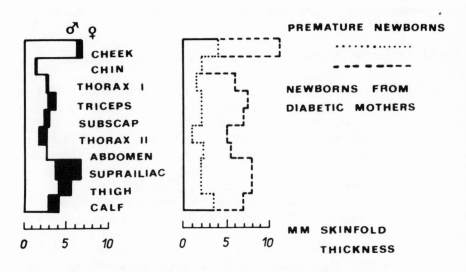

Figure 1. Mean values of fatfold thicknesses in 25
normal healthy boys (3450 g, s.d. = 477 g; 50.0 cm,
s.d. = 2.7 cm) and 23 girls (3287 g, s.d. = 439 g;
50.2 cm, s.d. = 1.0 cm) and in 23 premature children
and 18 overweight newborns from diabetic, metabolically
decompensated mothers (with hyperglycemia). The measurements were made between 4 and 48 hours after delivery by
a caliper.[8,9]

size (eight rats), weaned at the same age (30th day) and
kept in identical conditions with regard to nutrition
(Larsen diet ad libitum). At the age of 285-360 days,
the animals differed significantly in final weight and
body composition. These differences were dependent on
the degree of motor activity. The increased work load
group ran daily on a treadmill for one to three hours.
The group with restricted activity were in cages 8 x 12 x
20 cm. However, there was still considerable variability
within the subgroups (e.g., percent fat C.V. = 20 - 25).
Differences occurred in the final results despite uniformity in nutritional intake and exercise after
birth.[14-17] One possible factor is the number of litters
born from the mothers. These findings contribute to the
hypothesis concerning critical periods during development
when the organism is markedly more sensible to various
stimuli, especially nutritional, in regard to immediate
effects and future programming of body size and the
development of body tissues. Later influences cannot
always completely alter previously established trends.

THE IMPACT OF CALORIC BALANCE AND ENERGY TURNOVER

Another aspect concerning nutrition that has received
too little attention concerns not only the quantity of
food accepted, and its balance with caloric output, but
the level of both of them, that is, whether energy turnover is at a high or a low level. The overall metabolic
situation and its later impact on body composition and
body size, must be different depending on whether the
organism has a high or a low level of energy turnover.[18]
This is partly demonstrated by comparing a young growing
organism with an old one.[19] During growth, the organism
needs a great amount of calories for building body tissues.
Therefore the caloric input, relative to body weight, is
high (Figure 2). Simultaneously, spontaneous physical
activity is increased, with the peak around puberty[20] and
the basal energy output is very high also. Consequently,
the energy turnover is increased.[19] Simultaneously, the
oxidative processes and the activities of various enzymes
are at high levels. Even when lipogenesis is increased
during growth, as compared with later periods of life,[21]
the proportion of fat is low in a normal young organism.

An old organism is characterized by decreased caloric
input and output (low basal energy output, low spontaneous motor activity) and low tissue pO_2.[22] The proportion
of depot fat is increased in old experimental animals
(Figure 3).[16-19]   Corresponding findings have been reported from human subjects of advanced age (Figure 4)
even when subjects of the same relative weight are compared.[9,19,23]

Changes in the proportion of adipose tissue during
aging[24] are accompanied by the changes in its metabolic
activity, which is increased not only in experimental
animals, e.g., rats [15,17,19,25,26] but also in human
subjects.[27] Increased metabolic activity in young adipose
tissue is paralleled by increased ability to utilize fat

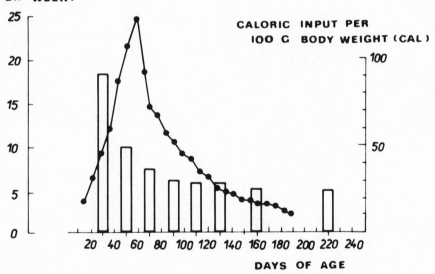

Figure 2.   Changes in the level of spontaneous physical
activity, measured in a revolving cage, and caloric
input relative to body weight, in male rats.[8,19,20]
The bars refer to caloric input.

in skeletal muscles.[17,19,28-30] Due to increased turnover in adipose tissue and increased ability to utilize
fat in muscles, a lower proportion of fat is kept in the
organism despite increased lipogenesis.[21]

   Further experiments showed that a greater degree of
myocardial lesions and a higher mortality after the application of isoprenaline was usual in old rats compared with
young rats. Aged rats with a low weight and low percentage of fat survived better than younger rats with high
proportion of fat.[31,32] This is related closely to body
size and composition (Figure 5).[33]

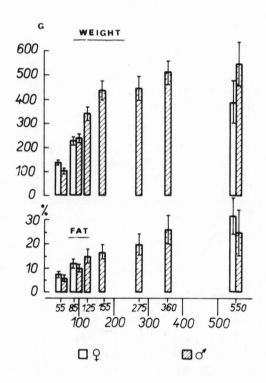

Figure 3. Changes in body weight and percent fat in
male and female rats (mean values and s.d.; 10-15 rats
in each group). The figures on the abscissa refer to
age in days.

METABOLIC AND MORPHOLOGICAL CONSEQUENCES

OF HYPOKINESIA DURING DEVELOPMENT

A low level of physical activity is usual in advanced
countries. The abolition of hard physical work has enabled people to develop their creative intellectual powers
but, obviously, the change was too sudden. Also, the
role of afferent stimulation of the central nervous system should not be underestimated.[34] Many attempt to
balance the lack of muscular work by decreased caloric
input to prevent an increase in weight and depot fat.

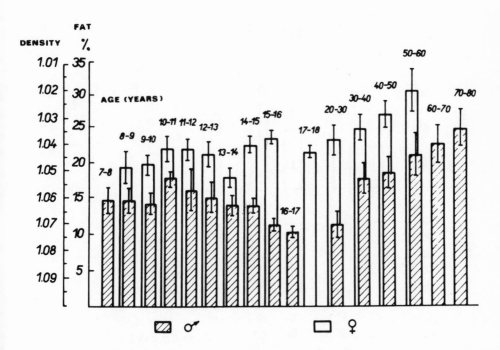

Figure 4. Changes in body density and percent fat in
human males and females from 7 to 70 years. Only subjects with normal, average weight according to age and
height (i.e., the same relative weight) were selected
for individual age categories. Each group includes
15-18 subjects (total males = 235; total females = 218;
means and s.e. are given).

As a result, they achieve the situation that is usual in
an old organism--hypokinesia and a reduced input of
calories (Figure 2). A very common finding in population
studies is a state of "hidden obesity"--an increased
amount of fat despite normal or even decreased body
weight.[10,15,26,35,36] This makes hypokinetic people
with small intakes similar to old subjects. It is necessary to stress also that many develop obesity that is
apparent.

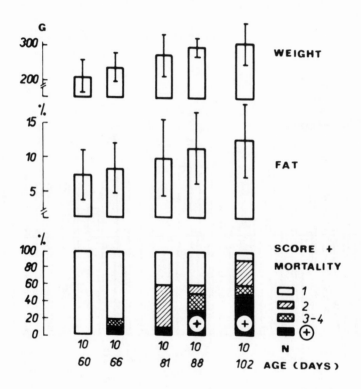

Figure 5. Weight, percent fat, the degree of experimental myocardial necrosis (evaluated by scores of 1 to 4
according to Rona et al.[31-33]) and spontaneous mortality
(+) after the administration of isoprenaline to male
rats of different ages.

There are more reasons why attempts should be made to keep the level of energy turnover on a level close to values found during earlier periods of development. These conclusions, based on data obtained from human subjects in Novosibirsk (USSR) living in nearly identical conditions, correspond generally to those from experimental animals. The highest level of motor activity occurred in the eight-year-old children when this was measured as the number of paces per day (mean = 16,200; range = 6,500 to 24,000).[37] People over 40 years of age, especially in more qualified posts, displayed much less activity (mean = 9,200; range = 3,500 to 25,000).[38]

Unfortunately data on adolescents were not reported, although this is the period when the highest level of activity could be expected. This was indicated, for example, by the results of our longitudinal study based on questionnaires of boys who were 11 to 18 years old.[39-42] Furthermore, a remarkable individual variability is apparent. In experimental animals, there are ten-fold differences in the levels of spontaneous activity.[24] Nevertheless, levels for individuals are relatively constant during longer periods but daily fluctuations occur. These tend to be characteristic for the individual. Each person shows "motor individuality." This is one reason why subjects with similar caloric input can differ in weight and body fat.

As indicated by developmental trends, a change in the level of physical activity ought to have different impacts at various stages of development. Experiments with laboratory animals showed that adaptation to hypokinesia (restriction of activity) was most marked during growth. It was manifest in an increased amount of body fat, decreased metabolic activity of adipose tissue, and a decreased ability of muscles to utilize fat metabolites.[14,17,26,33,43] When hypokinetic male rats aged 90 days were injected with labelled palmitic acid C^{14}, its turnover rate in the organism was slowed. The greater part of the palmitic acid was deposited in adipose tissue, and a lesser amount in the muscles (measured in soleus).[44]

THE IMPACT OF EXERCISE

Adult animals exercising daily on a treadmill deposited significantly smaller amounts of fat than control animals (Figure 2). The adipose tissue of these animals contained a greater proportion of desoxyribonucleic acid. That is, it was more cellular and therefore more like the adipose tissue of younger rats. Also, it was more active metabolically.[14,15,17,19,43,45] The turnover of injected palmitic acid C^{14} was more rapid, and its greater part was shifted to skeletal muscle in animals adapted, until adult, to an increased work load, compared with hypokinetic animals.[44]

The level of fat metabolism, usual in a young organism or after adaptation to increased physical activity, is linked closely to the aerobic capacity of the organism and the availability of O_2. A high pO_2 in the tissues reduces the changes in the ratio NADP/NADPH.[46] As a result, the reduction of dihydroxyaceton-phosphate, resulting in the accumulation of α-glycerophosphate, can increase to correspond with the accumulation of released free fatty acids (FFA). This allows their reesterification. The turnover of FFA in adipose tissue is increased and the fat content of the organism is kept low.[18] As mentioned above, a young organism is characterized by a high level of aerobic capacity relative to either total body weight or lean body weight;[47,48] hypoxia is usual in an old organism.[22,48] An increased aerobic capacity and the ability to deliver sufficient oxygen to the tissues is a basic characteristic of a trained organism.[41,42,47] This is true at all ages, especially in active sports[42] and is associated with an increased proportion of lean body mass at the expense of fat.[10,15,18,26,35]

It can be concluded, reasonably, that the organism that has adapted to systematic work load resembles the young organism in its physical activity regimen, body composition, aerobic capacity, selected characteristics of fat metabolism and in caloric input (Table I). In this respect, conclusions from experimental studies are in agreement with those from sportsmen, many of whom have exceptionally high caloric intakes (5000-6000 Cal/day)

but remain lean.[18,49] As is apparent from Table I, the
animals that were adapted to hypokinesia had the lowest
caloric input and, in this respect, they were similar to
old animals (Figure 2) but, despite this, they had larger
percentages of body fat than the exercised animals.[18]

The data obtained from the experimental animals with
different levels of physical activity suggest that
metabolic changes occurred. Some of these are reflected
in alterations of body composition that, at least in the
experimental animals, had a significant relationship to
the development of myocardial lesions after isoprenaline
was administered.[31,33]

The animals that were systematically exercised
(hyperkinetic) not only had a lower weight and lower proportion of fat but also had a lower incidence of myocardial
lesions and lower mortality after the same dose of isoprenaline (Figure 6). These findings are similar to those
in younger animals[32] (Figure 5) and are in agreement with
clinical observations.

TABLE I

Body weight, fat percent and caloric input in three
groups of rats differing in their amounts of exercise

Variables	Hyperkinetic		Control		Hypokinetic	
	$\bar{X}$	s.d.	$\bar{X}$	s.d.	$\bar{X}$	s.d.
Weight (g)	379.7	39.7	475.3	61.1	407.6	59.7
Fat (%)	8.5	1.8	19.4	3.1	14.5	3.9
Caloric input (g Larsen diet/100 g weight/day)	7.4	0.9	5.4	0.6	4.7	0.5

TRENDS IN GROWTH, BODY COMPOSITION AND

FUNCTIONAL CAPACITY IN CHILDREN

Conclusions derived from experimental animals and
from human subjects are in general agreement. Cross-
sectional and longitudinal studies of human subjects,
in many respects, however, allow more appropriate analyses
of changes in caloric input and overall energy turnover
and their impact on body size, body composition, functional capacity and fitness. A high level of physical
activity during growth and development is characteristic
for a normally and properly fed child only. Kraut[50]
claimed that the observation of children's play and its
liveliness and vivacity could provide good preliminary
information about children's nutritional status. In
advanced countries in which nutrition is commonly too
abundant and activity is restricted, other problems
appear. Some consequences, suggested both by experimental

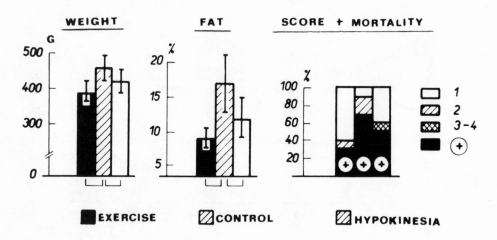

Figure 6. Weight, fat percent, the degree of experimental myocardial necrosis (scores of 1 to 4) and spontaneous mortality (+) after the administration of isoprenaline to male rats adapted to different levels of
physical activity[31] (mean values and s.d. are given;
12 to 15 rats in each experimental group).

data and common experience, could be expected. As
mentioned previously, e.g., the pathogenesis of atherosclerosis is now considered a pediatric problem.[4]
Enhanced deposition of depot fat during early development,
decreased rates of lipid metabolism and low aerobic
capacity, are each related to the levels of physical
activity.[18] Surely it is necessary to keep in mind the
remarkable individual variability in spontaneous physical
activity.[37,38] This could cause differential effects of
early nutritional manipulations among individuals.
Nevertheless, especially during early development, a low
energy turnover combined with a low level of physical
activity could contribute to the development of patho-
genetic states.

The effort to achieve optimal child development in
regard to body size, body composition, physical fitness
and health requires, in advanced countries, special
attention to rational nutrition and to the level of
physical activity. Satisfactory self-regulation of motor
activity at an optimal level cannot be taken for granted.
A low level of physical activity and muscular work,
together with reduced motor stimulation, is unsuitable
and illogical at a time when nutrition is more abundant
and body size is increasing.[18]

There appear to be significant differences between
the development of children who undergo regular systematic
exercise and the development of those who are relatively
inactive, i.e., only participate in physical education
at school. This is known from findings in a longitudinal
study of 41 boys from 11 to 18 years. The boys with the
highest levels of physical activity (more than six hours
intensive exercise per week) had significantly higher
relative and absolute amounts of lean body mass, at the
expense of fat, starting with the second to the third
year of the investigation (Table II).[18,39,40] This was
paralleled by increased aerobic capacity, ascertained as
the maximal oxygen uptake, both absolute and relative to
lean body mass, during increasing work load on a horizontal treadmill.[41,42] Moreover, body build differed among
groups separated according to their levels of physical
activity. The relative pelvic breadth (pelvic breadth/
body height or pelvic breadth/biacromial breadth) was

TABLE   II

Data from Boys Studied Longitudinally.
Group I--More Than 6 Hours Intensive Exercise Per Week.
Group III--No Organized Exercise except
Physical Education at School

	Group		1961	1962	1963	1964
Age	I.	$\bar{X}$	10.54	11.43	12.35	13.37
	(N=10)	SE	0.18	0.13	0.16	0.16
	III.	$\bar{X}$	10.70	11.80	12.38	13.38
	(N=13)	SE	0.16	0.14	0.14	0.14
Height	I.	$\bar{X}$	147.4	152.3	159.5	167.4
		SE	1.3	1.4	2.0	2.5
	III.	$\bar{X}$	143.9	149.2	154.5	161.8
		SE	1.4	1.5	1.6	1.8
Weight	I.	$\bar{X}$	37.6	40.7	46.7	53.4
		SE	1.8	1.8	2.3	3.1
	III.	$\bar{X}$	36.5	40.5	44.5	50.8
		SE	0.9	1.2	1.4	2.0
Lean body mass (%)	I.	$\bar{X}$	86.2	86.1	88.6	90.6
		SE	1.9	2.5	1.9	1.3
	III.	$\bar{X}$	83.2	81.9	83.2	85.8
		SE	1.3	1.8	1.5	2.1
Fat (kg)	I.	$\bar{X}$	5.3	5.7	5.2	5.0
		SE	0.9	1.1	0.9	0.8
	III.	$\bar{X}$	6.2	7.4	7.6	7.3
		SE	0.6	0.8	0.8	1.3
(Bicristal breadth/ height) x 100	I.	$\bar{X}$	13.5	14.5	14.3	13.6
		SE	0.2	0.2	0.2	0.3
	III.	$\bar{X}$	13.6	14.8	14.3	14.3
		SE	0.2	0.2	0.2	0.2

TABLE II (continued)

	Group		1965	1966	1967	1968
Age	I.	$\bar{X}$	14.26	15.58	16.60	17.68
	(N=10)	SE	0.12	0.15	0.15	0.12
	III.	$\bar{X}$	14.30	15.71	16.80	17.78
	(N=13)	SE	0.13	0.12	0.10	0.12
Height	I.	$\bar{X}$	173.5	179.2	182.2	183.3
		SE	2.4	1.5	1.1	1.2
	III.	$\bar{X}$	168.9	174.5	176.5	178.2
		SE	1.8	1.8	2.0	2.0
Weight	I.	$\bar{X}$	61.0	66.9	71.8	74.6
		SE	3.6	8.7	5.9	2.3
	III.	$\bar{X}$	57.8	64.0	67.5	69.2
		SE	1.9	1.7	1.4	1.6
Lean body mass (%)	I.	$\bar{X}$	91.9	91.5	90.8	92.4
		SE	1.0	1.4	1.0	2.8
	III.	$\bar{X}$	85.1	85.0	88.1	89.4
		SE	1.3	2.5	2.4	1.7
Fat (kg)	I.	$\bar{X}$	5.0	5.9	6.7	5.7
		SE	0.7	0.9	0.7	1.1
	III.	$\bar{X}$	8.6	9.6	8.6	7.2
		SE	0.9	1.1	0.4	1.4
(Bicristal breadth/ height) x 100	I.	$\bar{X}$	14.7	15.5	15.6	15.2
		SE	0.5	0.1	0.1	0.2
	III.	$\bar{X}$	15.9	15.6	15.9	15.9
		SE	0.2	0.3	0.2	0.2

significantly lower in most active boys, compared with
the least active ones. The boys in the latter group
tended to have large percentages of body fat, low aerobic
capacities and large relative pelvic breadths.[39,40,42]
The differences between the most and least active groups
(groups I and III) were most marked at the end of puberty
(14 to 15 years) when the differences in physical activity
were greatest. After this period, the levels of physical
activity tended to decrease in all three groups but the
differences in levels among the groups persisted. Significant differences in motor activity among the groups were
therefore preserved.[40-42]

Because there were some differences between the
groups at the age of 11 years, for example, a greater
height in group I, latent primordial variations in con-
stitution or "motor individuality" influencing interest
in exercise cannot be excluded even though these are not
manifest until later.[18] However, an analysis of covariance did not reveal any significant relationships between
the growth of lean body mass and of height.[18]

THE IMPACT OF OBESITY ON FUNCTIONAL CAPACITY

IN CHILDHOOD AND ADOLESCENCE

The interrelationships between nutrition, body size
and body composition can be demonstrated further using
obese subjects as an example. About 10 to 15 percent of
the children in advanced countries are obese. Obesity
is fairly common also in children of upper socioeconomic
classes in developing countries, e.g., Tunisia.[51] In
the development of obesity in childhood, a more important
role has been assigned to hypokinesia than to simple over-
eating.[18,52] An influence of the level of nutrition during
early development is strongly suggested.[6,7] However, a
multiple factor causation of the children's obesity should
be considered, including caloric inbalance and a low rate
of energy turnover.[18] An obese child is characterized
not only by increased relative and absolute amounts of
depot fat,[8] but also by an increase in lean body mass.[36]
Corresponding to the increase in body fat (Table III),
pelvic breadth is significantly increased compared with

normal boys.[40] It is stressed that these obese boys did
not display any endocrinological disturbances.[53] The
absolute values for maximal oxygen uptake were the same
as in normal boys the same age.[54] However, the oxygen
uptake values, relative to total or lean body mass, were
significantly lower indicating lowered aerobic capacity[41]
and, therefore, a lowered availability to the tissues of
the oxygen necessary for the metabolism of fat.

TABLE III

Comparisons between normal and obese children

Variables		Boys		Girls	
		Normal	Obese	Normal	Obese
Height	$\bar{X}$	161.8	161.2	156.9	157.5
(cm)	s.d.	6.3	2.1	5.7	4.0
Weight	$\bar{X}$	50.4	68.9	50.4	88.9
(kg)	s.d.	6.7	6.9	10.7	10.2
Fat	$\bar{X}$	12.5	29.5	18.1	31.0
(%)	s.d.	6.9	3.2	6.0	3.7
ATH	$\bar{X}$	43.9	48.6	40.7	46.7
(kg)	s.d.	5.1	5.0	4.9	6.4
Bicristal	$\bar{X}$	22.8	27.7	26.8	27.4
diameter (cm)	s.d.	1.8	1.0	2.3	1.1
Chest	$\bar{X}$	76.6	88.3	79.2	95.1
circumference(cm)	s.d.	3.3	5.5	4.2	3.8
Arm	$\bar{X}$	23.1	27.0	14.3	29.6
circumference(cm)	s.d.	1.2	2.3	2.5	3.3
Femoral condyle	$\bar{X}$	9.6	10.1	8.6	10.0
breadth (cm)	s.d.	0.3	0.9	0.4	0.5

This decreased aerobic capacity is paralleled by
low physical fitness and low work efficiency in the obese.
For example, the results of the step-test or a standard
work load on a bicycle ergometer indicate markedly increased energy cost for the same physical activity and
greater demands on the cardiorespiratory system,[46,53,55]
compared with normal or lean children. This is also one
reason for a further decrease in spontaneous motor activity in obese children. Even when these children participate in exercise or games, their activity is much lower
than for normal children. This has been demonstrated by
kinematographic methods.[52]

The reduction in fat and body weight, by adaptation
to increased exercise and work load, is paralleled therefore by improvements in work efficiency, physical performance and fitness.[53,55-57] Having achieved the same level
of maximal oxygen uptake after weight reduction, the
children were able to run longer on a treadmill at a
greater speed. That is, with the same aerobic capacity,
they performed better and longer than before weight
reduction.[41,56] Marked changes in the blood levels of
free fatty acids, during maximal work load, indicate an
increased ability to mobilize and utilize fat metabolites
as fuel for muscle work associated with the decrease in
body fat.[57,58]

THE EFFECT OF SYSTEMATIC EXERCISE

IN ADULTHOOD AND OLD AGE

In these groups, there are obvious interrelationships
among body build, body composition and physical fitness
in different states of nutrition and energy turnover. The
older the subject, the more difficult the differentiation
between primordial constitutional and hereditary factors
and those resulting from adaptation to stimuli during
development. There are spectacular examples in sportsmen,
especially those performing at championship levels. For
example, runners, as representatives of an active sport
discipline, have low absolute and relative body weights,
high proportions of lean body mass, relatively narrow
pelves[59] and high aerobic capacities.[60] In static sport

disciplines, e.g., weight lifting, sportsmen have high relative weights, high absolute amounts of lean body mass but also of depot fat, and relatively low aerobic capacities.[60] Some of these differences are due to a primary selection of particular body types for certain kinds of sport performance.[59] Mainly, this selection concerns total body height, length measures, and general body shape. The adaptation to intensive activity seems more important in body composition and aerobic capacity. Even body build, i.e., trunk breadths relative to body height, etc., could be modified markedly especially if training begins during childhood or adolescence (Table II). One obvious example is the Soviet Olympic team of girl gymnasts with their narrow pelves relative to body height or shoulder breadth.[61] Adaptation to increased work load is seen spectacularly in the bones of throwers.[62] The further analysis of constitutional and adaptive factors that influence the adult physique of champion sportsmen requires more experimental serial data, starting at early ages.

This differentiation between constitutional and adaptive factors appears in old people. Men who exercised regularly, during their whole life, displayed, in the 6th - 8th decade, greater amounts of lean body mass, greater circumferences of the forearm and thigh, lower fatfold thicknesses, higher aerobic capacities (Figures 7-8), greater muscle force and better performance in selected sports.[63-65] All the values were nearly the same as in control untrained, relatively inactive men ten years younger. This was apparent from cross-sectional and longitudinal studies.[63,64,66] Because the initial state of the exercised and control men is unknown, we cannot define the contribution of genetic factors as opposed to adaptation early in the development or later. Nevertheless, positive results of intensive life-long physical activity are obvious.

CRITERIA EVALUATING THE GROWTH AND NUTRITIONAL

STATUS OF CHILDREN

The preceding summary has directed attention to a

number of problems. In advanced countries, it is difficult, especially in large towns, to assure rational
nutrition and proper motor activity experience and to
achieve an optimum balance between these. This is reflected mostly in the development of body physique and
functional capacity in children, with possible consequences later in development. Moreover, technical progress and civilization are extending to developing
countries, together with the negative characteristics of
living styles in advanced countries. Measurements of
children in Tunisia confirmed that children from families
of the upper socioeconomic group, who have living

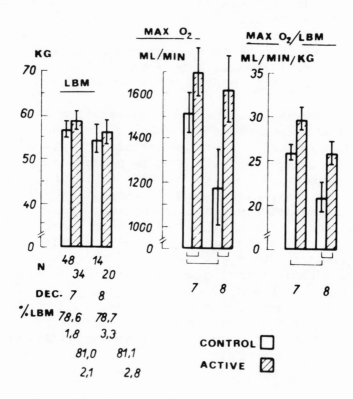

Figure 7. Lean body mass and aerobic capacity (evaluated from maximal oxygen consumption on a bicycle ergometer[66]) in active, exercised and inactive, control men
in seventh and eighth decades of life.

conditions similar to children in advanced countries,
display similar tendencies in their development--
accelerated somatic development, common obesity, etc.
A range of living conditions and economic levels enabled
some interesting comparisons. Boys from poor families
(but not underfed) had poorer somatic development, i.e.,
lower values of height, weight, lean body mass, etc.,
but higher functional capacities. The latter were shown
by more favorable cardiovascular reactions during work,
quicker recovery after work, greater relative muscle

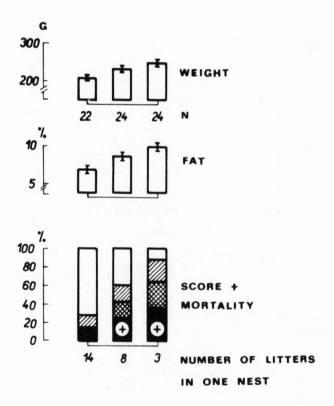

Figure 8. Weight, fat percent, the degree of experimental
myocardial necrosis (scores of 1 to 4) and spontaneous
mortality (+) after the administration of isoprenaline
to male rats nine to ten weeks old with different nutrition during early development.[32]

strength and better sport performance (Group A; Table IV) than boys of the same age from upper socioeconomic families (Group B). The latter had accelerated somatic development, but were retarded functionally.[51]

Further comparison of Tunisian girls aged 12 years, differing in the degree of sexual maturation, showed that the early maturers had more advanced somatic development (greater body height, weight, lean body mass, fat, more robust skeleton, broader pelvis and shoulders), but poorer reaction during a step test, slower recovery after it, relatively smaller muscle force and poorer efficiency in sports. Generally, functional development was retarded compared with the increases of body size and mass. Morphological acceleration was not associated with a functional advantage.[67]

TABLE   IV

Comparisons between Poor (Group A) and Upper
Socioeconomic (Group B) Tunisian Boys

Variables	Group A (N=30)		Group B (N=29)	
	Mean	s.d.	Mean	s.d.
Weight (kg)	31.4	4.4	36.8	7.1
Height (cm)	136.7	6.0	142.2	6.3
Chest circumference (cm)	70.8	3.4	75.4	5.8
Biacromial diameter (cm)	28.8	1.8	30.8	1.5
Bicristal diameter (cm)	21.7	1.3	22.9	1.5
Depot fat (%)	16.1	4.0	16.5	6.8
Lean body mass (kg)	26.3	3.0	30.5	3.9
Step-test index	118.9	13.4	111.1	15.8
Muscle strength (kp/kg/weight)				
--- trunk extensors	1.97	0.35	1.60	0.29
--- knee flexors	0.54	0.15	0.43	0.10
Broad jump (cm)	255.6	29.8	226.0	41.0

It is necessary to know how these children will look later in their development. Nevertheless, there appears a question whether the overall trends of acceleration, increase in body size and weight resulting from changes in nutrition, physical activity, etc., in advanced countries is desirable and positive when its present and late effects are considered. Experiences in developing countries could provide more experimental data, and contribute to defining the optimum in nutrition, motor stimulation and work load with variations in the balance between these throughout the life span but with special attention to the earliest stages of development.

Data from experimental animals show that, when comparing male rats weaned in litters of 3, 8 or 14, those that were least nourished during weaning (i.e., 14 in each litter) achieved the lowest total body weight, fat free body weight and percent fat at the age of 9 to 10 weeks. When these animals were compared in an induced pathological situation, i.e., the development of experimental cardiac necrosis after the administration of isoprenaline, the smallest animals were clearly at an advantage (Figure 8).[32]

The data reviewed indicate that a new evaluation of present tendencies in child care is necessary. Further knowledge is required of the associations between body size and body composition during development and the prognosis for functional capacity and health in adulthood and old age. The aim is to achieve the full potential of the organism. Not only undernutrition, overstraining and lack of care, but contrary factors in industrially developed countries, could be undesirable for human development.

REFERENCES

1.   Lewin, T.: Lecture to the Anthropological Society, Czech. Acad. Sci., Prague, 1970.

2.   Suchý, J.: Trend tělesného vývoje české mládeže ve 20. století. Čas.lék.česk., 110:935, 1971.

3.   Espenschade, A.S. and Meleney, H.E.: Motor
      performances of adolescent boys and girls of today
      in comparison with those of 24 years ago. Res.
      Quart., 32:186, 1961.

4.   Kannel, W.B. and Dawber, T.R.: Atherosclerosis as
      a pediatric problem. J. Pediat., 80:544, 1972.

5.   McCance, R.A. and Widdowson, E.M.: Nutrition and
      growth. Proc. Roy. Soc. (Biol.), London, 156:326,
      1962.

6.   Knittle, J.L. and Hirsch, J.: Effect of early
      nutrition on the development of epididymal fat
      pads: Cellularity and metabolism. J. Clin.
      Invest., 47:2091, 1968.

7.   Brook, C.G.D.: Evidence for a sensitive period in
      adipose-cell replication in man. Lancet, ii:624,
      1972.

8.   Pařízková, J.: Effect sur la croissance du jeune
      rat d'un régime librement choisi comparé à un
      régime gras. Nutr. Dieta (Basel), 3:236, 1961.

9.   Pařízková, J.: Age trends in fat in normal and
      obese children. J. appl. Physiol., 16:173, 1961.

10.   Pařízková, J.: The impact of age, diet, and exercise
      on man's body composition. Ann. N.Y. Acad. Sci.,
      110:661, 1963.

11.   Hoet, J.J.: Environmental factors in the etiology
      of diabetes mellitus. Proc. XIIIth Internat.
      Congr. Pediatrics. Metabolism, Vienna, Verlag
      Wiener Med. Akad., 1:397, 1971.

12.   Přibylová, H., Znamenáček, K. and Uedra, B.:
      Prenatal and postnatal influencing of early
      adaptation of children of diabetic mothers.
      Proc. XIIIth Internat. Congr. Pediatrics.
      Metabolism, Vienna, Verlag Wiener Med. Akad.,
      1:451, 1971.

13. Widdowson, E.M.: Early nutrition and later development. CIBA Foundation, Study Group No. 17, London, Churchill, 3, 1964.

14. Pařízková, J.: Physical activity and body composition. Proc. Conference on Body Composition, Soc. Study Human Biol., Oxford, Pergamon Press, 161, 1965.

15. Pařízková, J.: Nutrition, body fat and physical fitness. Borden's Rev. Nutr. Res., 29:41, 1968.

16. Pařízková, J.: Nutrition, related to body composition in exercise. Symposium Nutritional Aspects of Physical Performance. The Hague, Mouton, 89, 1972.

17. Pařízková, J.: Interrelationships of physical activity and nutrition to body composition and fitness, p. 85. In: Debry, G. and Blayer, R. (Eds) Alimentation et Travail. Premier Symposium International. Paris, Masson et Cie, 1972.

18. Pařízková, J.: Složení těla a lipidový metabolismus za různého pohybového režimu (Body composition and lipid metabolism in different regimens of physical activity). Prague, Avicenum, Hálek's Coll. 17:1, 1973.

19. Pařízková, J.: Compositional growth in relation to metabolic activity. Proc. XIIth International Congress of Pediatrics, Mexico, I:32, 1968.

20. Smith, L.C. and Dugal, L.P.: Age and spontaneous running activity of male rats. Canad. J. Physiol. Pharmacol., 43:852, 1965.

21. Gellhorn, A., Benjamin, W. and Wagner, M.: The in vitro incorporation of acetate-1-C^{14} into individual fatty acids of adipose tissue from young and old rats. J. Lipid Res., 3:314, 1962.

22. Sirotinin, N.N.: Hypoxia at elderly and senile age. Proc. 9th Internat. Congr. Gerontology, Kiew, 1:254, 1972.

23. Pařízková, J.: Sledování rozvoje aktivní hmoty u dospívající mládeže metodou hydrostatického vážení. Čs. Fysiol., 8:426, 1959.

24. Brožek, J. and Keys, A.: Evaluation of leanness-fatness in man: Norms and interrelationships. Brit. J. Nutr., 5:194, 1951.

25. Altschuler, H., Lieberson, M. and Spitzer, J.J.: Effect of body weight on free fatty acids release by adipose tissue in vitro. Experientia, 18:91, 1962.

26. Pařízková, J.: Nutrition and its relation to body composition in exercise. Proc. Nutr. Soc., 25:93, 1966.

27. Novák, M., Monkus, E. and Pardo, V.: Metabolism of the subcutaneous adipose tissue in human newborns in the early neonatal period. Proc. XIIIth Internat. Congr. Pediatrics, Metabolism, Vienna, Verlag Wiener Med. Akad., 1:67, 1971.

28. Pařízková, J. and Koutecký, Z.: The effect of age and different motor activity on fat content, lipoprotein-lipase activity and relative weight of internal organs, heart and skeletal muscle. Physiol. Bohemoslov., 17:177, 1968.

29. Pařízková, J., Koutecký, Z. and Staňková, L.: Fat content and lipoproteinase activity in muscles of male rats with increased or reduced motor activity. Physiol. Bohemoslov., 15:237, 1966.

30. Beatty, G. and Bocek, R.M.: Metabolism of palmitate by fetal, neonatal, and adult muscle of the rhesus monkey. Amer. J. Physiol., 219:1311, 1970.

31. Pařízková, J. and Faltová, E.: Physical activity, body fat and experimental cardiac necrosis.
    a) Brit. J. Nutr., 24:3, 1970. and
    b) Physiol. Bohemoslov., 18:503, 1969.

32. Faltová, E., and Pařízková, J.: The effect of age, body weight and body fat on experimental myocardial necrosis. Physiol. Bohemoslov., 19:275, 1970.

33. Pařízková, J.: The effect of age and various motor activity on fat content, lipoproteinase activity and experimental necrosis in the rat heart. Med. and Sport, 3:137, 1969.

34. Smirnov, K.M.: Gipokinez i obraz zhizni tchelovyeka. In: Dvigatelnaya aktivnost tchelovyeka i gipokineziya. Akad. Nauk. SSSR, Novosibirsk, 11, 1972.

35. Pařízková, J.: Body composition and physical fitness. Curr. Anthrop., 9:273, 1968.

36. Pařízková, J.: Activity, obesity and growth. Monog. Soc. Res. Child Develop., No. 140, 35:28, 1970.

37. Ledovskaya, N.M.: Dvigatelnaya aktivnost doshkolnikov v usloviyakh g. Novosibirska. In: Dvigatelnaya aktivnost tcheloveka i gipokinesiya. Akad. Nauk SSSR, Novosibirsk, 22, 1972.

38. Gapon, A.Ja.: Issledovanija povsyednevnoy dvigatelnoy aktivnosti u litz umstvenvennogo truda. In: Dvigatelnaya aktivnost teheloveka i ginokinesiya. Akad. Nauk SSSR, Novosibirsk, 46, 1972.

39. Pařízková, J.: Longitudinal study of the development of body composition and body build in boys of various physical activity. Human Biol., 40:212, 1968.

40. Pařízková, J.: Somatic development and body composi-
    tion changes in adolescent boys differing in
    physical activity and fitness: A longitudinal
    study. Anthropologie, 10:3, 1972.

41. Pařízková, J. and Šprynarová, Š.: Developmental
    changes in boys build, composition and functional
    aerobic capacity in boys. Symposium Energy
    Expenditure and Physical Activity. Nutrition –
    Proceedings of the VIIIth International Congress,
    Prague. Excerpta Medica, 316, 1970.

42. Šprynarová, Š.: Dynamika pohybových podnětů se
    zřetelem k rozvoji tělesné zdatnosti mládeže.
    Proc. IInd Internat. Congr. Phys. Fitness Youth,
    Prague, 374, 1966.

43. Pařízková, J. and Staňková, L.: Release of free
    fatty acids from adipose tissue in vitro after
    adrenaline in relation to the total body fat in
    rats of different age and different physical
    activity. Nutr. Dieta (Basel), 9:43, 1967.

44. Poledne, R. and Pařízková, J.: Long-term training
    and net transport of plasma free fatty acids.
    Proc. 2nd International Symposium on Exercise
    Biochemistry, Magglingen. Basel, Karger, 1973.

45. Pařízková, J. and Staňková, L.: Influence of physical
    activity on a treadmill on the metabolism of adi-
    pose tissue in rats. Brit. J. Nutr., 18:325, 1964.

46. Paul, P.: Uptake and oxidation of substrates in the
    intact animal during exercise, p. 225. In:
    Pernow, B. and Saltin, B. (Eds) Muscle Metabolism
    during Exercise. Advances in Experimental Medicine
    and Biology Ser. New York, Plenum Press, 1971.

47. Åstrand, P.-O.: Human physical fitness with special
    reference to sex and age. Physiol. Rev., 36:307,
    1956.

48. Pařízková, J., Šprynarová, Š., Macková, E. and Eiselt, E.: Changes in aerobic capacity as related to lean body mass in the ontogeny of man. Physiol. Bohemoslov., 22:425, 1972.

49. Pařízková, J. and Poupa, O.: Some metabolic consequences of adaptation to muscular work. Brit. J. Nutr., 17:341, 1963.

50. Kraut, H.: Food intake as a factor of production, p. 215. In: Debry, G. and Blayer, R. (Eds) Alimentation et Travail. Premier Symposium International, Vitell. Paris, Masson et Cie, 1972.

51. Pařízková, J. and Merhautová, J.: The comparison of somatic development, body composition and functional characteristics in Tunisian and Czech boys of 11 and 12 years. Human Biol., 42:391, 1970.

52. Mayer, J. (Ed): Overweight: Causes, Cost and Control. Englewood Cliffs, N.J., Prentice Hall, 1968.

53. Pařízková, J., Vaněčková, M., Šprynarová, Š. and Vamberová, M.: Body composition and fitness in obese children before and after special treatment. Acta. Paediat. Scand., Suppl., 217, 1971.

54. Šprynarová, Š. and Pařízková, J.: Changes in the aerobic capacity and body composition in obese boys after reduction. J. appl. Physiol., 20:934, 1965.

55. Pařízková, J., Vaněčková, M. and Vamberová, M.: A study of changes in some functional indicators following reduction of excessive fat in obese children. Physiol. Bohemoslov., 11:351, 1962.

56. Pařízková, J.: Obesity and physical activity. Proc. Nutrition Symposium, Nutritional Aspects of Physical Performance. The Hague, Mouton, 1972.

57. Pařízková, J. and Vamberová, M.:   Body composition
     as a criterion of the suitability of the reducing
     regimens in obese children. Develop. Med. Child
     Neurol., 9:202, 1967.

58. Pařízková, J., Staňková, L., Šprynarová, Š. and
     Vamberová, M.:   Influence de l'exercice physique
     sur certains index métaboliques sanguins chez les
     garcons obèses après l'effort.
     a) Summ. Čs. Fysiol., 12:392, 1961. and
     b) Nutr. Dieta (Basel), 7:21, 1965.

59. Pařízková, J.:   Masse active, la graisse déposée
     et la constitution du corps chez les sportifs
     du haut niveau. Kinanthropologie, 4:95, 1972.

60. Šprynarová, Š. and Pařízkovă, J.:   Functional
     capacity and body composition in top weight-
     lifters, swimmers, runners and skiers. Int. Z.
     angew. Physiol., 29:184, 1971.

61. Pařízková, J. and Horná, J.:   Unpublished data.

62. Kuratchenkov, A.I.:   Travmatism yavleniya treniro-
     vannosti i netrenirovannosti kostno-sustavnovo
     apparata u metateley. Trudy LNIIFK 5, Leningrad,
     5: 1949.

63. Pařízková, J. and Eiselt, E.:   Body composition and
     anthropometric indicators in old age and the
     influence of physical exercise. Human Biol.,
     38:351, 1966.

64. Pařízková, J. and Eiselt, E.:   A further study of
     changes in somatic characteristics and body com-
     position of old men followed longitudinally for
     8-10 years. Human Biol., 43:318, 1971.

65. Kuta, I., Pařízková, J. and Dýcka, J.:   Muscle
     strength and lean body mass in old men of different
     physical activity. J. appl. Physiol., 29:168, 1970.

66. Fischer, A., Pařízková, J. and Roth, Z.: The effect of systematic physical activity on maximal performance and functional capacity in senescent men. Int. Z. angew. Physiol. Arbeitsphysiol., 21:269, 1965.

67. Merhautová, J. and Pařízková, J.: Functional capacity and somatic development in children of Tunis. Proc. XVIIIth World Congress of Sports Medicine. Oxford, 1970.

FROM A QUAC STICK TO A COMPOSITIONAL ASSESSMENT OF MAN'S NUTRITIONAL STATUS

Josef Brožek

From Lehigh University, Bethlehem,
Pennsylvania, USA

Let us begin with a brief, selective bibliographic review. The recommendations for the anthropometric appraisal of nutritional status, formulated by the enlarged Committee on Nutritionl Anthropometry, of the (U.S.) National Research Council, were published in 1956 in Human Biology[1] and as an introduction to a volume entitled Body Measurements and the Evaluation of Nutritional Status.[2]

The history and the theoretical bases of the in vivo analysis of body composition were presented in a large paper, written jointly with Ancel Keys, and published in Physiological Reviews, in 1953, under the title "Body Fat in Man."[3] A detailed account of quantitative body composition models appeared in the American Journal of Physical Anthropology,[4] the journal that carried the first paper on the estimation of tissue masses on the basis of anthropometric data.[5] The topic was examined in a more comprehensive way elsewhere, taking into consideration models of body composition, methods, and illustrative applications; 163 references were cited.[6]

For much of the technical detail, the interested individual still must turn to Techniques for Measuring Body Composition, published in 1961 by the NRC-NAS in Washington and reprinted subsequently as a governmental publication.[7] The techniques of measurement are discussed also in a two-volume opus, entitled Body Composition,[8] and in a "sister publication," entitled Human Body Composition that reported the proceedings of a conference held in London, attended by European scientists.[9] However, the focus of these two publications is on the application of the techniques and on the results that were obtained.

In a paper published 20 years ago, I stressed the mutual importance of physical anthropology and nutritional science, the one providing tools, the other topics, problems and opportunities for the development of a dynamically oriented physical anthropology.[10] This point, I believe, still holds true.

ORGANISM, PHYSIQUE, AND NUTRITIONAL STATUS

We may begin, in earnest, by some simple-minded, approximative modeling. The organism will be represented (Figure 1) by a circle (which may be interpreted as the letter "O," for "organism"--the totality of the organism's characteristics). The organism is embedded in a matrix of environments--geological (E_1), atmospheric (E_2), socio-economic (E_3), cultural (E_4) and nutritional (E_5).

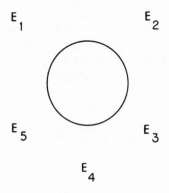

Figure 1. The organism in a matrix of environments.

Some of the organism's characteristics are affected
by the supply of nutrients (the diet) and their ingestion,
absorption, and assimilation. We may refer to them as
"N" characteristics, denoting "nutritional status." The
remaining characteristics may be labeled "R," the
"remainder" (Figure 2).

The number of the "N" characteristics and their
identity will vary widely in different dietary conditions.
This can be illustrated by using two extremes: vitamin A
deficiency and prolonged caloric deficiency, in initially
normal young men. In the first case, only a few characteristics will be involved, if we work with adults in a
controlled, laboratory situation. These are the level of
vitamin A in blood plasma, the altered photochemistry of
the cones and especially of the rods, and, on the functional level, impaired dark adaptation. The "N"--slice
of the circle is very thin (Figure 3). We could separate
it into a chemical and a "functional" (physiologico-
psychological) sector. By contrast, in prolonged calorie
deprivation many characteristics will be affected (including, importantly, the bulk and the composition of the
body). The "N"--characteristics occupy a large part of
the circle; the "R"--slice (the "remainder") is, consequently, thin (Figure 4).

Furthermore, different characteristics will be
affected to different degrees. Thus, at the functional
level, the speed of hand movements is affected but little

Figure 2. A diagrammatic representation of the characteristics of an organism. A = nutritional characteristics;
R = remainder.

in semi-starvation but strength is impaired profoundly, and endurance in hard physical work drops precipitously. At the morphological level, bone mineral will decrease little, if at all, while protein stores will be affected substantially and fat stores will decrease dramatically.

We shall illustrate the relative magnitude of the changes by considering alterations in gross body weight and in total body fat, estimated densitometrically, in a specific study.[11] The gross body weight of the subjects decreased in six months of semi-starvation to about 75 percent of the control value. There was a return toward the pre-experimental value, first at a slow rate (during restricted refeeding, R1-R12), then more rapidly, substantially exceeding the control figures, and then slowly returning toward the "normal" (see Table I).

These decreases and increases were much more dramatic for body fat, estimated densitometrically. At R58 body fat was still somewhat elevated, while the gross body weight was almost back to the control, pre-starvation value. Gross body weight would have been an inadequate indicator of nutritional status.

Now we are ready to consider the complex of external body measurements--the very core of our considerations. When we think of them in reference to a normal state of the organism, we can represent them as a biconvex segment extending both into the N-region and the R-region (Figure 5). In other words, some of the A-measures ("A" for

Figure 3. A diagrammatic representation of vitamin A deficiency. N = nutritional characteristics; R = remainder.

anthropometry) are relevant for the characterization of
nutritional status while others fall outside the N-area.

Now let's go from the general model to the two specific examples of changes in the organism induced by
altered diet. In simple vitamin A deficiency, all the
body measurements are outside the N area (Figure 6).
They are not affected by this dietary situation. Consequently, they are irrelevant (and would be inappropriate)
as measures of "nutritional status." By contrast, very
many--though not all--anthropometric characteristics
change under the impact of prolonged negative calorie
balance (Figure 7).

THE FIVE CENTRAL THESES

I shall condense my further comments into five
theses. They are presented as clearly visible targets
at which to shoot.

1. Deficient supply or impaired utilization of
    nutrients affects various facets of the organ-
    ism: morphological, biochemical, physiological,
    behavioral. The characterization of "nutritional
    status" (and, especially, of the changes in
    nutritional status) must take into account a
    multiplicity of variables, if the subject is
    to be covered adequately. Altered morphology
    is only one facet of "nutritional status" and

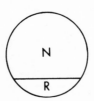

Figure 4. A diagrammatic representation of prolonged
calorie deficiency. N = nutritional characteristics;
R = remainder.

its importance will vary in different dietary
conditions.

2. Even when we focus on the morphological aspects
    (as we are doing, largely, at this conference)
    we still have to deal with a complex system that
    includes both the external morphology and, even
    more importantly, the internal morphology of
    tissue masses.

3. The quantitative description of the morphologi-
    cal aspects of nutritional status can range from
    one or more external body measurements to a
    multi-parameter system of body components.

4. Our choice of measurements will depend on the
    nature of the dietary deficiency, on our aims,
    and on the facilities available.

TABLE I

Body Weight (W%) and Estimated Body Fat (F%)
in Semistarvation and Nutritional
Rehabilitation, as Percentages of the
Control Value [11]

Period	C	S24	R12	R20	R33	R58
N	32	32	32	20	21	6
W%	100	75.8	84.7	104.9	109.3	103.0
F%	100	34.0	67.2	151.7	139.1	109.8

C = control period, S = weeks of semistarvation,
R = weeks of rehabilitation, N = the number
of subjects on which the means are based.

5. Models of body composition imply certain quantitative assumptions. When these assumptions are
   not satisfied (e.g., due to the presence of
   edema), estimates of body composition based on
   simplified, "standard" models, will be inaccurate.

THE VEXING ISSUE OF EDEMA

We encountered edema as a complicating factor in
the very first collaborative study of body composition
carried out in a nutritional context.[11] Edema was measured as excess thiocyanate space (excess above the control values). The thiocyanate space values are larger
than the true extracellular space, but we were concerned
with the relative, rather than the absolute values,
because we wished to follow the changes taking place in
the course of semi-starvation and subsequent refeeding.

In semi-starvation, the thiocyanate space is expanded
(Table II). However, when we express the experimental
values of the "extracellular" space as percentages of
the control value, the true magnitude of the excess hydration is underrepresented, because the absolute <u>increase</u>
in the volume of the extracellular fluid is accompanied
by a simultaneous, marked <u>decrease</u> in total body weight.
The second row in Table II gives values of the fluid
equivalent of the thiocyanate space expressed as percentages of body weight. The changes in semi-starvation are,
clearly, dramatic.

Figure 5. Relationship of anthropometric measures (A)
to nutritional characteristics (N) and the remainder (R).

Smith used tritium to measure total body water in
24 malnourished infants under the age of two years. Upon
admission, the mean total body water represented 84.5
percent of body weight. After recovery, the value fell
to 62.6 percent.[12] This excessive hydration, typically
present in malnourished children, masks the loss of
tissues.

THE LIMITATIONS OF NUTRITIONAL ANTHROPOMETRY

The figures presented above indicate, in as clear
a fashion as one might wish, the limitations of external
body measurements as a basis for the characterization
of the somatic aspects of nutritional status. The amount
of edema cannot be readily estimated "anthropometrically."
Furthermore, the edema must be substantial before its
presence can be established somatoscopically and clinically. Finally, the presence of edema alters the quantitative assumptions built into the simple equations for
the estimation of body fat (and of its complement, the
"lean" body mass) on the basis of body density.

These statements should not be interpreted as an
attack on either nutritional anthropometry or on the
densitometric approach to the assessment of body composition. All of us here, including this speaker, are in
favor of the use of body measurements for the characterization of nutritional status. At the same time, it is

Figure 6. Representation of vitamin A deficiency showing
that all anthropometric measures (A) are within the area
for the remainder (R) and outside that for nutritional
characteristics (N).

the specific responsibility of the enthusiast to point
out the limitations of his methodology.

MEASURING NUTRITURE: SIMPLE INDICATORS

The title of the present paper implies a continuum
of procedures, increasing in sophistication and in the
complexity of measurement. Elsewhere with special reference to body composition, I referred to a series of
"successive approximations."[13]

We would probably agree that mid-arm circumference
is the most promising candidate for the title of the
simplest anthropometric criterion of nutritional status,
especially of undernutrition. In a recent paper,
Bradfield et al. restated the view that "Midarm circumference can indeed be used as a rapid and cheap indicator
of severe protein-calorie malnutrition," adding, importantly, "if measurements can be compared with standards
for children of the same age."[14] The trouble is, as the
authors note, that in many areas parents do not know the
precise age of the children. Several approaches have
been proposed to compensate for the frequent absence of
reliable information regarding the child's age. Thus
Kanawati and McLaren have suggested the use of the head
circumference as a reference point, and calculated ratios
of mid-arm circumference to head circumference.[15]

Figure 7. Representation of prolonged calorie deficiency.
Most anthropometric measures (A) are within the area for
nutritional characteristics (N).

The "QUAC stick" belongs to the same category of
simple procedures but utilizes body height as the point
of reference.[16] QUAC is an abbreviation of "Quaker Arm
Circumference." The procedure was developed by a British
Quaker relief team operating an emergency relief program
in Nigeria in 1968/69.[17] The mechanics of assessment
is simplicity itself. A strip with a height scale and
corresponding selected values of the upper arm circumferences is attached to a flat vertical surface. Children
whose upper arm circumference measures below a value regarded as critical (e.g., 80 or 85 percent of the "normal"
value), are identified readily. The cut-off values are
printed on the strip used to measure the child's height.
In evaluating these simple somatic indicators of "nutritional status" we must keep in mind the purpose for which
they were designed: the identification, under conditions
of emergency, of children most affected by insufficiency
of the food supplies.[16] They are useful in identifying
children below a certain cut-off point, not for fine
discrimination across the whole range of undernutrition-
overnutrition.

TABLE II

Thiocyanate dilution space (T%), as
percentage of the control value,
and the fluid equivalent (T),
expressed as percentage of body
weight determined at the
corresponding times [11]

Period	C	S24	R11	R19
T%	100.0	109.1	107.3	102.4
T	23.6	34.0	30.6	23.8

C = control period, S = weeks of
semi-starvation, R = weeks of
rehabilitation

The mid-arm circumference/height, just as weight/height, has the advantage of simplicity. For purposes of research we may wish to know not only the total body mass (or its sample, as in the case of mid-arm circumference) but its composition. Clearly, body weight, in the presence of substantial excess extracellular fluid, is not a dependable criterion of nutritional status. Upper arm circumference is likely to be less affected by excess hydration than the gross body weight, and for this reason if no other, it is preferable as a benchmark of undernutrition.

MEASURING NUTRITURE: BODY COMPOSITION

The presence of edema (even the possible presence of edema) forces us to move from traditional anthropometry to thinking in terms of body composition. The human body consists of components that respond differentially to calorie deficit (or surplus), the bony structures being affected least, (if at all), and the body fat showing the largest fluctuation.

At the theoretical level, applying the principle of "successive approximations," one possible sequence may involve, as the first step, the separation of the fat component from the "remainder" (R_1) which is the "fat free" body mass (Table III). The latter is relatively homogeneous, i.e., more homogeneous than total body weight. We consider body fat as a "body component" and the "fat free body mass" as a "body compartment" susceptible of further analysis. The next steps may involve determination of extracellular and intracellular water, and separation of "body solids" into body protein and body minerals, osseous and non-osseous.

The actual steps will depend on the purposes of the study, the available technology, and special considerations, including concern about disturbed body hydration. This may necessitate a simultaneous determination of several parameters, thus making the "linear" model in Table III inapplicable. For an approach stressing a simultaneous analysis of multiple components of body composition, see Moore et al.[18]

SUMMARY

The system of compositional analysis (or, as I prefer
to call it, somatolysis) is versatile, both conceptually
and operationally, i.e., in the way the components and
compartments are defined and measured. It is a natural
and, in some situations, a necessary extension of classi-
cal body measurements.

In the area of somatometry, measurements of the fatfolds constitutes an important supplement to lengths,
diameters and circumferential measurements, and a step
towards thinking in terms of specific tissue masses rather
than in terms of complex, heterogeneous variables, such
as gross body weight or limb circumferences.

At times it is useful to remind ourselves of truisms.
One such truism is that morphological description of an
organism, no matter how sophisticated it may be, covers
only one aspect of "nutritional status." It does not tell
us the vitamin A concentration in the plasma nor how well
we could find our way around Burg Wartenstein, should
electricity fail and we had to depend on our adaptation
to the darkness.

TABLE III

Body Weight (W), Body Components, and
"Residual Masses" or Body Compartments
(R1-R4)

W = Body fat + R_1 [fat-free body mass]

R_1 = Extracellular water + R_2
[cell mass and extracellular minerals]

R_2 = Intracellular water + R_3 [solids]

R_3 = Body protein + R_4 [minerals]

R_4 = Osseous minerals + non-osseous minerals

REFERENCES

1. National Research Council, Committee on Nutritional Anthropometry: Recommendations concerning body measurements for the characterization of nutritional status. Human Biol., 28:111, 1956.

2. Brožek, J. (Ed): Body Measurements and Human Nutrition. Detroit, Wayne Univ. Press, 1956.

3. Keys, A. and Brožek, J.: Body fat in adult man. Physiol. Rev., 33:245, 1953.

4. Brožek, J.: Body composition: Models and estimation equations. Amer. J. Phys. Anthrop., 24: 239, 1966.

5. Matiegka, J.: The testing of physical efficiency. Amer. J. Phys. Anthrop., 4:223, 1921.

6. Brožek, J.: Human body composition--Models, methods, applications. Anthropologie (Brno), 3:3, 1966.

7. Brožek, J. and Henschel, A. (Eds): Techniques for Measuring Body Composition. Washington, D.C., Nat. Acad. Sci.- Nat. Research Council, 1961. (Also: Government Res. Report AD 286506, Office of Technical Services, U.S. Department of Commerce, 1963).

8. Brožek, J. (Ed): Body composition. Annals N.Y. Acad. Sci., 110:1, 1963.

9. Brožek, J. (Ed): Human Body Composition. Oxford, Pergamon Press, 1965.

10. Brožek, J.: Measuring nutriture. Amer. J. Phys. Anthrop., 11:147, 1953.

11. Keys, A., Brožek, J., Henschel, A., Mickelson, O. and Taylor, H.L.: The Biology of Human Starvation. Minneapolis, University of Minnesota Press, 1950.

12.   Smith, R.:   Total body water in malnourished infants.
      Clin. Sci., 19:275, 1960.

13.   Brožek, J.: Foreword.   In: Meneely, G.R. and Linde,
      S.M. (Eds) Radioactivity in Man.   Second Symposium.
      Springfield, Ill., Charles C Thomas, 1965.

14.   Bradfield, R.B., Jelliffe, E.F.P. and Jelliffe, D.B.:
      Assessment of marginal malnutrition.   Nature,
      235:112, 1972.

15.   Kanawati, A.A. and McLaren, D.S.:   Assessment of
      marginal malnutrition.   Nature, 228:573, 1970.

16.   Gowrinath, S.J.:   Evaluation of QUAC-stick for
      growth assessment in children.   Indian J. Med.
      Res., 60:747, 1972.

17.   Arnhold, R.:   The QUAC stick: A field measure used
      by the Quaker Service team, Nigeria.   J. Trop.
      Pediat., 15:243, 1969.

18.   Moore, F.D., Olesen, K.H., McMurray, J.D., Parker,
      H.V., Ball, M.R. and Boyden, C.M.:   The Body Cell
      Mass and Its Supporting Environment.   Philadelphia,
      W.B. Saunders Co., 1963.

CHANGES IN PIGS DUE TO UNDERNUTRITION BEFORE BIRTH, AND FOR ONE, TWO, AND THREE YEARS AFTERWARDS, AND THE EFFECTS OF REHABILITATION

Elsie M. Widdowson*

From the Dunn Nutritional Laboratory, University of Cambridge and Medical Research Council, Cambridge, England

Studies on the relation between nutritional status and physical anthropometry in man are, of necessity, limited to measurements of those parts of the body that are accessible in the living person. Animals, on the other hand, can be killed, and the effects of alterations in nutritional status on the structure of the inner parts of their bodies can be investigated.

This paper describes studies we have made on pigs, first undernourished and growth retarded in utero, and second, severely undernourished for one, two and three years after birth. In each study, we have compared the undernourished animals with well-nourished controls, both of the same size but younger, and of the same age but much larger. Some of the pigs were rehabilitated after each phase of undernutrition, and the effects of this are described also. We have measured the dimensions of the pigs at the various stages, weighed their organs and

* Present address: Department of Investigative Medicine, University of Cambridge, Downing Site, Downing Place, Cambridge CB2 1QN, England

analyzed them. Some of the results are likely to be of
general application and to provide information about the
effect of undernutrition on the physical anthropometry
and body composition of children. Unusual body proportions or body composition in older children, or adults,
might be due to early nutritional experiences.

    In biological research, it is often helpful to exag-
gerate the conditions likely to apply in the human situation. The newborn pig weighs 1 to 1.5 kg as compared
to 3 to 3.5 kg for the human baby at birth. The adult
pig weighs about 300 kg so, in this species, weight at
birth is only 0.3 to 0.5 percent of the adult weight in
contrast to man where the corresponding value is five
percent. It is possible, theoretically, therefore to
retard the growth of a pig far more than the growth of
a human baby is ever likely to be retarded.

UNDERNUTRITION BEFORE BIRTH

    Nature sometimes makes the "Wigglesworth"[1] experiment
for us, and, in a large litter of pigs, one is born weighing only one-third to a half as much as its littermates.
This pig is small because the blastocyst was implanted
at an unfortunate site in the uterine horn, where the
blood flow was poorer than elsewhere. Table I shows the
mean weights of the bodies and organs of newborn "runt"
pigs from seven litters compared with those of seven
immature fetuses of the same body weight, taken at 90
days gestation, and with their own larger littermates
of the same age.[2] The last column expresses the values
for the runts as a percentage of those for their larger
littermates.

    A comparison of the first and third columns shows
the changes in weight that occur during normal development in utero over the last 25 days of gestation, during
which time the body increases in weight by about 2.5 times.
Liver, kidneys, heart, spleen, stomach and quadriceps
muscles increased in weight between two and three times,
and therefore grew approximately in parallel with the
body as a whole. The high weight of the lungs of the
fetuses was due to the fluid that they contain before it

is replaced by air immediately after birth.

It seems likely that the growth of the runt pigs
began to fall behind well before 90 days gestation,
because the weight of their muscles was significantly
less than that of muscles of the well grown fetuses of
the same weight or size, although their livers, kidneys
and small intestines were about the same size. Hearts,
spleens and stomachs were retarded in a parallel with the
body as a whole.

Table II gives values for the whole brain and for
the forebrain, cerebellum and brain stem, set out in the
same way as Table I.[3] The brains of the runt piglets
weighed less, but not much less, than those of their

TABLE I

Mean Weights (g) of Body and Organs of Seven Full
Term "Runt" Pigs and Their Larger Littermates, and
of Seven Immature Fetuses of the
Same Body Weights as the "Runts"

	Fetus	"Runt"	Larger litter-mate of "runt"	"Runt" as % of litter-mate
Body	619	625	1606	39
Small intestine	14.8	15.4	50.5	26
Liver	14.9	12.6	45.7	28
Quadriceps muscles	7.2	4.4	14.1	31
Lungs	21.3	11.1	31.4	36
Kidneys	4.4	4.0	10.1	40
Heart	4.0	5.0	11.8	43
Stomach	2.8	4.0	8.6	47
Spleen	0.8	1.1	2.1	53

larger littermates. The forebrain, which contributed
over 80 percent of the weight of the brain, did not grow
nearly as fast as the cerebellum over the last 25 days
of gestation. The cerebellum nearly trebled in weight,
and much of this growth was achieved by the runt piglet.
The brain stem is even more striking for it was the same
weight in the runt pig as in its larger littermate,
weighing 2.5 times as much.

The internal organs of the runt pigs contained less
protein and DNA than those of their larger littermates
and they appeared to have fewer and smaller cells. The
brains of the runts were more highly developed chemically
than those of the fetuses, but less highly developed as
a whole than those of their larger littermates.

We have also examined the bones and the same general
principles apply to them.[4] The bones of the runts were
between those of the fetuses and larger littermates in
size and development. However, they were nearer to the
larger littermates, and age seemed to be the major

TABLE   II

Weight of Brains (g) of Full Term "Runts" and
Larger Littermates and of Immature Fetuses

	Fetus	"Runt"	Larger litter-mate of "runt"	"Runt" as % of litter-mate
Whole brain	21.4	28.2	32.3	87
Forebrain	19.1	22.9	26.6	86
Cerebellum	1.3	2.9	3.3	88
Brain stem	1.0	2.4	2.4	100

determinant of the differential development of the
skeleton and its parts.

These are all particular examples of a general principle, namely that the effect of undernutrition is to
delay the growth of all tissues, but in descending order
according to their priority of growth at the particular
age in question.

GROWTH OF THE RUNT PIG AFTER BIRTH

Some runt pigs and their larger littermates have
been well fed and allowed to grow to maturity. The runts
never quite made good their deficiency in weight or
length. Table III shows the dimensions of two of these
animals when they were killed at three years of age after
they had stopped growing. The body of the runt was shorter
and its long bones were shorter and lighter than those
of its littermate that had been larger at birth. This
must be an important reason why all the other parts of
the body are smaller, for the growth of the skeleton
determines the ultimate size of the body. If the bones
stop growing before the soft tissues have reached their
full dimensions, these will only grow to a size appropriate to the skeleton that supports them. This is illustrated in Table IV, which shows the weights of the organs.
The muscles, liver and heart were all heavier in the
larger pigs and the organs of the larger pigs contained
more protein and more DNA (Table V). If the values for
the total amount of DNA in the organs at maturity are
compared with those for the runt pig and its littermate
at birth, it is clear that there has been approximately
a tenfold increase in both animals.

SEVERE UNDERNUTRITION AFTER BIRTH

We have given pigs so little food that they weigh
only five to six kg when they are a year old although
their well-nourished littermates weigh about 200 kg.
This enables us to compare, in the same species, the
effects of severe undernutrition before birth with that
of severe undernutrition afterwards. We start the

deprivation when the pigs are about 10 days old and weigh three to four kg. We give them just enough food to enable them to grow very slowly during the year. Table VI shows the weights of their organs and those of their littermates one year old. The starting point is the 10-day-old pig, and the weights of its organs are shown also. Some parts of the body have barely changed in weight during the year of undernutrition. Again, skeletal muscle is particularly retarded. It actually weighed less than it did a year earlier, and the liver and kidneys had not gained much weight. In these respects, therefore, undernutrition after birth produced the same effect as slow growth in utero. The small intestine more than doubled in weight and therefore grew more than the

TABLE III

Dimensions of Two Pigs at Three Years

	"Runt"	Large littermate
Weight at birth (g)	450	1500
Weight at 3 years (kg)	309	368
Length of "side" from articulation of femur to first thoracic vertebra(cm)	100	119
Length of "loin" from articulation of femur to last thoracic vertebra(cm)	49	62
Depth of "side" across widest part of chest (cm)	55	59
Measurements of long bones		
Femur---weight (g)	710	857
--- greatest length (cm)	27	30
--- width at midpoint (cm)	3.39	3.84
--- depth at midpoint (cm)	3.41	3.74
Tibia and fibula weight (g)	548	669
--- greatest length (cm)	25.2	28

body as a whole. In the runt piglet, the small intestine
grew less than the body. The intestine has no digestive
function before birth but there is very rapid growth
immediately afterwards in response to food.

Table VII gives values for the brain, which continued
to grow during the year of undernutrition, and more than
doubled its weight. The increase was mostly in protein,
lipids and water.[5] The brains of the undernourished animals weighed 67 percent as much as those of their well-
nourished littermates although their bodies weighed less
than three percent as much. The cerebellum was nearer
its correct adult weight than either the forebrain or the
stem. Undernutrition of this degree of severity after
birth hindered the normal development of the brain, so
that the brains remained chemically immature for the
chronological age. The forebrain, for example, contained
a lower percentage of cholesterol than the forebrain of
the well-nourished animals.

The small undernourished animals naturally had
shorter bones than their well-nourished littermates, but
more striking was the structure of the long bones. These

TABLE IV

Weights of Organs of
Three-Year-Old Pigs

	"Runt"	Large litter-mate of "runt"
Whole body (kg)	309	368
Liver (g)	1088	2613
Kidneys (g)	495	502
Heart (g)	718	883
Gastrocnemius muscle(g)	218	251

were wide for their length and the cortex was very thin
and dense, with the marrow cavity filled with dark red
gelatinous material.[6] The bone crystals were large and
the collagen highly calcified (Table VIII). Fractures
were not observed. The bones were adequate for the body
they had to support and, in the same way, the kidneys,
hearts, liver, lungs and other organs, small though they
were in the undernourished animals, were adequate for
the function they had to subserve. When food is short,
some parts of the body are affected more than others,
and those least affected are those whose normal function
is maintained by the metabolic processes vital for the
life of the animal. The body adapts itself to severe
shortage of food in such a way that the animal or child
survives.

The organs of the undernourished pigs contained much
less DNA than those of their large well-nourished littermates, and the same was true of skeletal muscle. Muscle
fibers are multinucleate and, although the measurement of
DNA gives information about the number of nuclei, it gives
no indication of the number of muscle fibers. In the pig
and human baby, the full number of muscle fibers is

TABLE   V

DNA (mg) in the Organs of "Runt" Pigs and
Their Larger Littermates at Birth and
When Three Years Old

| | Birth | | Three years | |
	Runt	Larger litter-mate	Runt	Larger litter-mate
Liver	99	169	1055	2000
Kidneys	39	69	1320	1450
Heart	29	46	314	407

acquired by about the time of full term birth.[7,8,9] The severe degree of undernutrition to which these pigs were subjected after birth made no difference to the number of fibers in their muscles and the small undernourished pig had as many as the large one.[10] The size of the fibers, however, was very different in the two animals. The number of fibers is determined genetically before birth but, within any one litter of pigs, the number of fibers in a given muscle is similar whatever the size of the animal.

One striking difference in appearance between the undernourished pigs and normal pigs of the same body weight, was the size of the heads and ears relative to the rest of the body. This is illustrated in Table IX,

TABLE VI

Mean Weights of Body and Organs of Eighteen Under-nourished and Seven Well Nourished One-Year-Old Pigs, and of 10 Ten-Day-Old Well Nourished Pigs Representing the Start of the Experiment

	Well nourished 10 days	Under-nourished 1 year	Well nourished 1 year	Under-nourished as % well nourished 1 year
Body (kg)	3.4	5.7	214	2.7
Gastrocnemii(g)				
Muscles	22.5	13.0	635	2.0
Spleen (g)	6.9	6.7	276	2.4
Lungs (g)	50.8	62.4	1723	3.6
Heart (g)	22.2	35.7	566	6.3
Liver (g)	116.0	139.0	2198	6.4
Stomach (g)	16.5	67.7	938	7.3
Kidneys (g)	22.8	25.6	344	7.5
Small intestine (g)	93.7	220.0	1695	13.0

which shows that the external ears had grown considerably
during the year of undernutrition so that they were nearly
twice as long as those of younger animals of the same
body weight.

As time has gone by, we have become more adventurous
and we have continued the undernutrition beyond one year
to two years and even three years. We have had to allow
the animals to gain weight very slowly all the time.
The general effects on the body have remained the same.
The response to rehabilitation after longer periods of
undernutrition has proved to be the most interesting part
of the work.

RESPONSE TO REHABILITATION

When the pigs undernourished for one, two or three
years after birth are allowed access to plenty of food,
they at once begin to eat a great deal of it and to grow
very rapidly. Figure 1 shows the progress of three groups
of rehabilitating females undernourished for different

TABLE VII

Weight of Brain (g) of Undernourished
and Well Nourished Pigs

	Well nourished 10 days	Under- nourished 1 year	Well nourished 1 year	Under- nourished as % well nourished 1 year
Whole brain	39.3	81.5	125	65
Forebrain	31.5	63.9	98.7	64
Cerebellum	4.2	9.1	13.4	88
Brain stem	3.6	8.5	13.3	61

lengths of time compared with that of a normal animal,
Susan, who was in fact their mother. Her gain in weight
is typical of the growth of well-nourished females in our
herd.

All rehabilitating animals showed catch-up growth
at first, though the longer they had been undernourished
the less rapid was the gain during the initial period.
Well-nourished females stop gaining weight after three
years; those undernourished for one year gained weight
for two years and then stopped growing roughly at the
same age as the well-nourished animals, so that they
become smaller adults. Those undernourished for two years
gained weight for about 18 months and became smaller
adults still. When those undernourished for three years
began to be rehabilitated they had already reached the
age when growth normally stops. They, too, grew for a
time, but they stopped gaining weight when this was only
about half that of the well-nourished pigs. They were
small in every way and Table X illustrates this. It
shows the mean dimensions of four groups of female animals, well-nourished, and undernourished for one, two
or three years and then fully rehabilitated. The longer
the undernutrition lasted, the length of the body and

TABLE   VIII

The Effect of Undernutrition
on the Calcium/Collagen Ratio
in the Cortex of the Pig's Humerus

Undernourished animals	1.63
Animals of the same size (but much younger)	1.16
Animals of the same age (but much larger)	1.39

limbs was successively smaller, but the fatter the animals
became. These standard measurements of the thickness of
the subcutaneous fat layer do not give a true picture of
how fat the animals became because the muscles, particularly those of the buttocks, in the animals rehabilitated
after two and three years undernutrition, were so infil-
trated with fat that the muscle fibers were completely
embedded within it.

     This raises questions regarding current ideas about
the development of fat cells. It has been suggested that
overfeeding in infancy causes rapid multiplication of
fat cells which in turn leads to an excessive number of
fat cells in the adult and consequent obesity.[11-14]
This may be so, but the findings on the pigs suggest that
the converse may also be true. These animals had cells
full of fat at 10 days of age when undernutrition began
but these cells became completely empty and remained so

TABLE   IX

Measurements of Undernourished Pigs One Year Old and of
Well Nourished Pigs of Similar Body Weight

	Undernourished	Well nourished of the same weight
Body weight (kg)	6.10	5.27
Length of head (cm)	19.8	14.0
Length of body (cm)	46.4	37.2
Width across head between ears (cm)	7.7	5.3
Greatest length of ears (cm)	11.3	6.3
Greatest width of ears(cm)	10.8	7.9

for the whole of the period of undernutrition, whether
it lasted one, two or three years. The pigs began to
deposit fat rapidly in their bodies as soon as plentiful
food was supplied, and the longer the period of deprivation the fatter they tended to become.

Those who have had the care of marasmic children
agree that such children often become fat when they are
rehabilitated. We have no information yet about the
number of fat cells in marasmic children before and after
rehabilitation. If Brook's[14] hypothesis is correct, that
the first year after birth is the sensitive period for
fat cell multiplication in man, then one would expect
children severely undernourished before one year to
develop less than the average final number of fat cells.
They may do, but, like the pigs, they lay down fat rapidly
when they are rehabilitated. We have still not solved
all problems concerning the effects of early nutrition

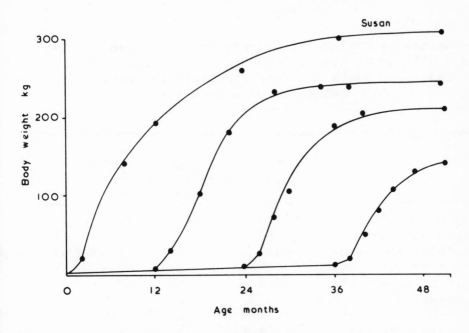

Figure 1. Gain in weight of pigs undernourished for
one, two or three years and then rehabilitated, compared
with a pig well nourished throughout.

on the deposition of fat in the body, and on the number
of cells produced to accommodate it.[15]

The weights of the organs were appropriate for the
size of the body in the rehabilitated animals (Table XI)
and none of them reached the size of those of animals that
have never been undernourished. None of the organs contained as much DNA as the corresponding organs of the
well-nourished animals, or of animals undernourished for
a shorter period. There was a large increase in DNA in
all organs, however, and that is perhaps more striking
than the fact that nuclear division stopped too soon.

The pigs do not mature sexually while they are so
severely undernourished, though some development of
sexual organs goes on. These mature very rapidly as
soon as plentiful food is supplied and males and females
mate and the females produce litters when they are a very

TABLE X

Dimensions of Well Nourished Female Pigs Three to Five
Years Old and of Female Pigs Undernourished for One,
Two and Three Years and Then Rehabilitated Until They
Ceased to Gain Weight

	Well nourished	Period of Undernutrition		
		1 year	2 years	3 years
Body weight (kg)	290	220	193	127
Length of hind leg (cm)	80	68	68	64
Length of side (cm)	111	101	99	85
Length of loin (cm)	52	48	47	35
Depth of side (cm)	52	48	48	45
Thickness of back fat:				
Shoulder (mm)	6.9	6.5	6.5	6.5
Mid back (mm)	4.2	3.3	3.5	3.0
Loin (mm)	5.0	4.4	5.2	5.5

small size. We have had litters from females undernourished for one, two and even three years. The size of
each newborn piglet is about average, but the number in
the litter tends to be small. These piglets grow normally
if they are well fed and become as large as their grand-
mothers and other well-nourished animals. Two daughters
of a pig undernourished for three years, who herself
weighed 127 kg when she was fully grown, weighed over
200 kg when they were 19 months old. They are still
with us, and still growing.

TABLE XI

Weights of Organs of Well Nourished Pigs Three to
Five Years Old, and of Pigs Undernourished for One,
Two and Three Years and Then Rehabilitated Until
They Ceased to Gain Weight

	Well nourished	Period of undernutrition		
		1 year	2 years	3 years
Body (kg)	290	220	193	127
Liver (g)	2954	2125	1671	1397
Kidneys (g)	580	444	322	262
Heart (g)	736	572	448	400
Brain (g)	140	121	111	111

REFERENCES

1.   Wigglesworth, J.S.: Experimental growth retarda-
      tion in the foetal rat. J. Path. Bact., 88:1,
      1964.

2.   Widdowson, E.M.: Intra-uterine growth retardation
      in the pig. I. Organ size and cellular develop-
      ment at birth and after growth to maturity. Biol.
      Neonate, 19:329, 1971.

3.   Dickerson, J.W.T., Merat, A. and Widdowson, E.M.:
      Intra-uterine growth retardation in the pig.
      III. The chemical structure of the brain. Biol.
      Neonate, 19:354, 1971.

4.   Adams, P.H.: Intra-uterine growth retardation in
      the pig. II. Development of the skeleton. Biol.
      Neonate, 19:341, 1971.

5.   Dickerson, J.W.T., Dobbing, J. and McCance, R.A.:
      The effect of undernutrition on the postnatal
      development of the brain and cord in pigs. Proc.
      Roy. Soc. B., 166:396, 1967.

6.   Dickerson, J.W.T. and McCance, R.A.: Severe under-
      nutrition in growing and adult animals. 8. The
      dimensions and chemistry of the long bones.
      Brit. J. Nutr., 15:567, 1961.

7.   Stickland, N.C.: Ph.D. Thesis, University of Hull,
      1973.

8.   MacCallum, J.B.: On the histogenesis of striated
      muscle fibers and the growth of the human sarto-
      rius muscle. Bull. Johns Hopkins Hosp., 9:208,
      1898.

9.   Montgomery, R.D.: Growth of human striated muscle.
      Nature, 195:194, 1962.

10. Stickland, N.C., Widdowson, E.M. and Goldspink, G.: Effects of low protein and low calorie diets on the number of fibres and nuclei in the skeletal muscle. (In preparation)

11. Hirsch, J. and Knittle, J.L.: Cellularity of obese and nonobese human adipose tissue. Fed. Proc., 29:1516, 1970.

12. Knittle, J.L.: Childhood obesity. Bull. N.Y. Acad. Med., 47:579, 1971.

13. Brook, C.G.D., Lloyd, J.K. and Wolf, O.H.: Relation between age of onset of obesity and size and number of adipose cells. Brit. Med. J., ii:25, 1972.

14. Brook, C.G.D.: Evidence for a sensitive period in adipose-cell replication in man. Lancet, ii:624, 1972.

15. Widdowson, E.M. and Shaw, W.T.: Full and empty fat cells. Lancet, ii:905, 1973.

EFFECT OF MATERNAL DIETARY PROTEIN ON

ANTHROPOMETRIC AND BEHAVIORAL DEVELOPMENT OF THE OFFSPRING*

Bacon F. Chow

From the Department of Biochemical and Bio-
physical Sciences, The Johns Hopkins University
School of Hygiene and Public Health, Baltimore,
Maryland, USA

Anthropology is a science that treats of the growth
of human beings, among other topics. Growth of human
beings deals with not only physical measurements such as
height, weight, head circumference, fatfold thickness,
fat content, etc., but should include weights and functions of organs. Above all, the metabolism of the whole
human being should be included. In our laboratory, we
have recognized since 1942 that individuals with equivalent height and weight may have organs of different size,
function and metabolism. Further, we have demonstrated
that the composition of tissues, such as blood and
organs, can be altered through the diet, and, more
recently, we have demonstrated that the metabolic function of animals and humans can be altered in utero,
particularly through the diet of the mother. With this
in mind, we present herewith the three parts of our
paper.

ANIMAL EXPERIMENTS

First, we will deal with animal experiments. It

* After the untimely death of Dr. Chow, the final preparation of this paper was completed by his close associ-
  ate, Andie M. Hsueh, Sc.D.

has been found in our laboratory that dietary restriction
during gestation and lactation results in undesirable
effects in the offspring including: high mortality in
early life, low birth weight, permanent growth stunting,
delayed physical development, feed wastage, and behavioral
abnormalities despite ad libitum feeding of an adequate
stock diet after weaning.[1-6] In this paper, we will
report further studies in the identification of (1) the
critical period in development affected by dietary
restriction, i.e., gestation and/or lactation; (2) the
critical dietary component, i.e., protein, calories or
other, the lack of which will result in disadvantaged
pups; and (3) a specific organ, the underdevelopment of
which causes the above cited damages.

Experimental Methods

Female rats from the colony are raised until the
age of five to six months on Purina Laboratory Chow fed
ad libitum. Sets of six females, each of body weight
240-260 g are mated overnight with two males of body
weight 400-425 g. Mating is continued each night until
sperm are seen in the vaginal lavage that is examined
each morning. When the females become sperm-positive,
they are housed individually and assigned randomly to
different dietary regimes. After weaning, all the
progeny are housed individually and fed an adequate
stock diet on an unlimited basis.

The following animal models were prepared to determine the effect of dietary restriction of the dams during
gestation and/or lactation on the offspring.

     RR -- pups from restricted mothers and nursed by
           other restricted mothers,

     RN -- pups from restricted mothers nursed by
           ad libitum fed mothers,

     NR -- pups from ad libitum fed mothers nursed by
           restricted mothers, and

NN -- pups from ad libitum fed mothers nursed by
    other ad libitum fed mothers.

The restricted mothers receive 10 g Purina Laboratory
Chow daily during pregnancy and 20 g for the first 21 days
of lactation; corresponding to approximately 50 percent
of the ad libitum consumption of normal rats during gestation and lactation respectively.

The following groups of progeny were prepared to
demonstrate the importance of protein in the maternal
diet:

CC -- progeny of dams fed 20 percent casein diet
    ad libitum during gestation and lactation,

CR -- progeny of dams fed 10 percent casein diet
    ad libitum during gestation and lactation,

WG -- progeny of dams fed 20 percent wheat gluten
    diet ad libitum during gestation and lacta-
    tion, and

RR + C -- progeny of dams fed a 40 percent casein
    diet but only half the intake of controls
    (CC) during gestation and lactation.

Purina Laboratory Chow was used in studies to determine the effects of overall maternal dietary restriction.
Formulated diets were used to determine the crucial components in the diets. There were four formulated diets.

(1) 20 percent casein diet is the basal one which
    has the following composition:

	g/kg diet
Casein	200
Sucrose	700
Alpha-cellulose	20
Mazola oil (corn oil)	40
Hegsted Salt Mixture IV	40
Vitamin diet fortification mixture	22
(Nutritional biochemical Corp.,	
Cleveland, Ohio)	

(2)   10 percent casein diet--replace 100 g of casein
       with 100 g of sucrose for each kg of 20 percent
       casein diet,

(3)   40 percent casein diet--replace 200 g of sucrose
       with 200 g of casein in each kg of 20 percent
       casein diet,

(4)   20 percent wheat gluten diet--replace 200 g of
       casein with 200 g of wheat gluten in each kg of
       20 percent casein diet.

Results

Figures 1a, 1b and 1c demonstrate that when dietary
restriction was imposed on the dams during pregnancy,
the progeny showed a delayed stunting effect that was not
as severe as that observed in the progeny from dams
restricted either during lactation alone or during both
gestation and lactation (Figure 1a). The data also show
that if the dams were given sufficient calories but either
a low quantity of a good protein or sufficient of a poor
quality protein (Figure 1b); or the dams were given a
sufficient amount of good quality protein but only half
the caloric intake of the normal dams (Figure 1c), the
progeny from all three types of dams were permanently
stunted in growth.

Two types of behavioral tests used on the above
progeny will be reported in this paper. One is the maze
running test and the other is the test for Conditioned
Emotional Response (CER). The detailed procedures of
both tests have been published.[7,8] Figure 2a shows the
results of maze running of NN and RN progeny and Figure 2b
shows the results of maze running of NN and NR progeny.[9]
They demonstrate the difference in learning behavior of
the progeny from dams restricted during gestation (RN)
or lactation (NR). RN rats not only took longer to start
running the maze, but they made more errors than the NN
rats. Even at the end of 20 trials, they did not reach
the error-free stage, while the NN rats became error-free
after the 15th trial. NR rats showed a slightly slower
starting time than the NN rats. However, they made

essentially the same number of errors as the NN rats.

Figures 3a, 3b and 3c show the results of the CER test of RN, RR + C and WG rats with their own control groups.[8,10,11] In these tests, rats were put into a cage with two compartments linked by a gate. Water deprivation for 24 hours preceded this test and water was placed in the second compartment as a reinforcement. Each rat was placed in the first compartment and had to learn to cross the gate to obtain water. In the eighth trial, a single electric shock was given via the water. The results showed that both RN (Figure 3a) and WG (Figure 3c) rats were slower to learn to cross the gate

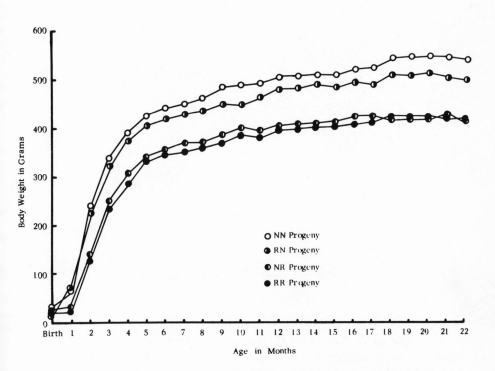

Figure 1a. Growth Curves of NN, RN, NR and RR Male Rats.[2]

and, after the shock, they were hesitant to go towards
the water.   Before receiving the shock, the RR + C rats
(Figure 3b) did not behave differently from the controls
(NN rats).   After the shock, RR + C rats were more
cautious than the controls, but, after the fifth trial
following shock, there was no significant difference
between the RR + C rats and the controls.   These results
demonstrate the importance of the maternal diet during
gestation as well as showing that the crucial component
in relation to behavior performance of the offspring is
protein, not calories.

     A number of observations suggested that the abnormal
behavior of the RN rats was related to an underdevelopment

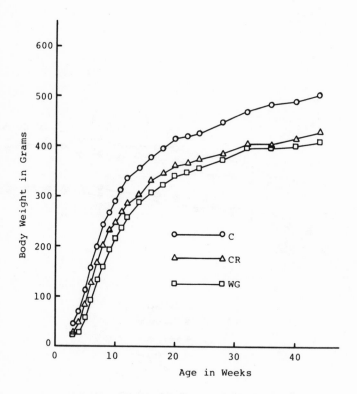

Figure 1b.   Growth Curves of Male Progeny from Mother
Rats Fed 20 percent Casein (C) or 10 percent Casein (CR)
or 20 percent Wheat Gluten (WG) Diet during Gestation
and Lactation.

of their pituitary glands rather than to a brain defect. Therefore, we determined the growth hormone content and activity of the pituitary gland of such animals.[12] The results demonstrate that both growth hormone activity and growth hormone content of the pituitary of the progeny from mother rats restricted during gestation and lactation periods were reduced significantly (Tables I and II). The smaller pituitaries of RR rats and their lower growth hormone content, accompanied with their abnormal behavior, led us to inject pituitary extract into the RN rats from the 7th to the 28th day of life. This produced normalization of the behavioral performance of RN rats in the CER test[13] (Figure 4).

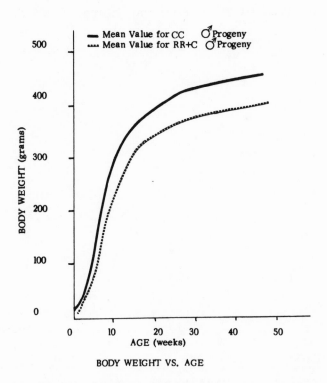

BODY WEIGHT VS. AGE

Figure 1c. Growth Curves of CC and RR+C Progeny. In this study, 1 g of dl-methionine was added to each 100 g of casein in the maternal diet.[10]

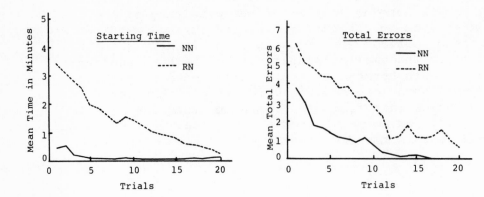

Figure 2a.   Results of Maze Running Tests in NN and RN
Progeny.[9]

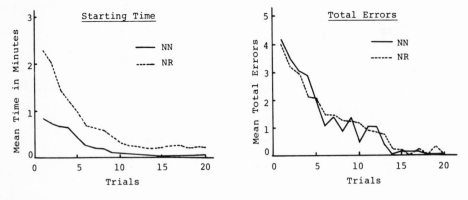

Figure 2b.   Results of Maze Running Tests in NN and NR
Progeny.[9]

These results provide evidence that restriction
during gestation brings about delayed stunting and
behavioral aberrations, whereas restriction during lactation only brings about permanent and severe growth
stunting, but no decrease in learning capability.

Our studies indicate also that both the anthropometric and behavioral damage observed in the offspring
of restricted dams is found in the offspring of dams
consuming either poor quality protein or good protein
in an insufficient amount.

In an attempt to find a common denominator that would
explain the metabolic, behavioral and physical damage to
the offspring, due to the inadequacy of maternal diet,
it was reasoned that the difficulties of the disadvantaged
pups might be due to changes in the pituitary rather than
in the brain or other organs. The pups of underfed dams
showed retarded development of this master gland as was
demonstrated by radioimmunoassay for growth hormone

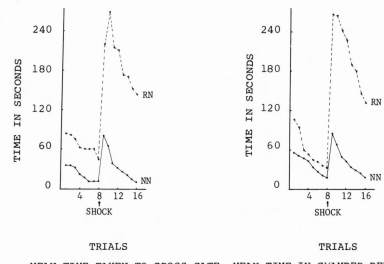

MEAN TIME TAKEN TO CROSS GATE   MEAN TIME IN CHAMBER BEFORE DRINKING

Figure 3a. Results of Conditioned Emotional Response
Tests in NN and RN Rats.[8]

content and bioassay for growth hormone activity. However, the final proof of the hypothesis that the pituitary
gland is probably the key organ that is damaged lies in
the fact that the injection of crude pituitary extract,
without the presence of growth hormone, can normalize the
behavioral performance of gestationally restricted pups.
The injection of growth hormone alone improves the growth
but not the behavioral damage.

HUMAN STUDIES

The results obtained from our rat experiments provided us only a lead as to how malnutrition of the dams
affected the development of the offspring. Our ultimate
goal is to utilize the knowledge obtained from rat studies
to better understand human beings. Once a rat model is
established, it becomes extremely useful in conducting
experiments in humans. After it was found that restriction of the total dietary intake of the dams during gestation and lactation produced unfavorable progeny as
measured by growth, metabolic and behavioral development,

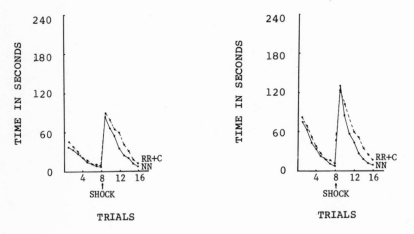

MEAN TIME TAKEN TO CROSS GATE   MEAN TIME IN CHAMBER BEFORE DRINKING

Figure 3b. Results of Conditioned Emotional Response
Tests in NN and RR+C Rats.[10]

it became important to test the same hypothesis in human
populations. After three years of preparatory work, a
longitudinal study began in January 1967. This study was
funded jointly by The Johns Hopkins University, U.S. Naval
Medical Research Unit No. 2, the Chinese Government (Joint
Commission on Rural Reconstruction and the National Science
Council), Rockefeller Foundation, the Ambrose Monell
Foundation, and the Agency for International Development.
The specific objective was to determine the effect of the
addition of a high protein supplement to the diets of
pregnant women, normally consuming a basic vegetable diet
of marginal protein intake, on the growth, food utiliza-
tion, neuromotor and mental development and susceptibility
to infection of the offspring. The site of the study is
Suilin township located in the western center of Taiwan
(Republic of China).

Women enrolled in this project had to meet the
following criteria:

1. Age 20-28 years,

2. Habitual diet estimated to supply less than 40 g
    of protein daily,

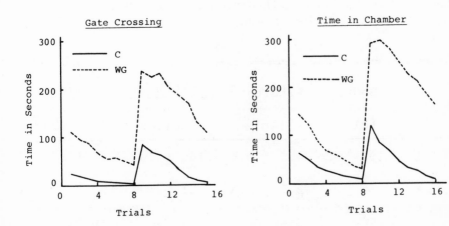

Figure 3c. Results of Conditioned Emotional Response
Tests in WG and CC Progeny.

3.   Married with at least one normal child,

4.   In third trimester of pregnancy at the time of enrollment,

5.   Intend to have more children as soon as possible,

6.   Fully cooperative, and

7.   Normal upon routine physical examination with hemoglobin not less than 11 g percent, hematocrit not less than 36 percent and plasma proteins not less than 5.5 g/100 ml.

The total number of pregnant women enrolled was 294. Two liquid supplements (A or B) of the same color and flavor were used. The subjects were assigned randomly to either supplement group. Supplement A contained 20 g of milk protein and 400 Kcal plus minerals and vitamins in each 12 ounce can. Supplement B contained no protein,

TABLE   I

Means and Standard Errors of Pituitary Growth Hormone Content (Radioimmunoassay) of Pups from Restricted (RR) and Ad Libitum Fed (NN) Dams (McCollum Strain, Six Weeks of Age, Males)

| Group | N | Body weight (g) | Pituitary weight (mg) | Growth hormone content | |
				µg/ pituitary	µg/mg pituitary
NN	6	153 ± 4	4.5 ± 0.4	149 ± 19	33 ± 4
RR	8	64 ± 6	2.7 ± 0.3	51 ± 10	20 ± 3
P		<0.001	<0.01	<0.001	<0.01

less than 40 Kcal and the same levels of minerals and
vitamins. Subjects began supplementation at three weeks
after delivery (this infant is called the first study
infant). Each subject consumed two 12 ounce cans of
either Supplement A or Supplement B daily. They drank
the liquid supplement directly from the can so that no
comparison between the supplements could be made by the
subjects. The subjects consumed their respective supplement in the presence of our field nurses. Thus any un-
consumed portions of the supplement could be measured and
recorded. However, all the subjects were encouraged to
consume the entire amount each time. The entire investigation was carried out as a double blind study. The
workers did not know who was consuming which supplement;
in fact, our field workers were told that there were more
than two kinds of supplement preparations.

The consumption of the supplements continued through
the lactation of the first study infant (approximately
15 months for lactation), the interim before the second

TABLE II

Effect of Dietary Restriction of the Dam on Growth
Hormone Activity of the Hypophysis of the Offspring

Source of pituitary gland	N	Mean increase in body weight (g) $\pm$ S.E.M.	Equivalence in terms of μg bovine growth hormone preparation per mg pituitary
NN	8	24 $\pm$ 0.3	22.0
RR	8	15 $\pm$ 0.6	5.1

$$P < 0.001$$

pregnancy, during the second pregnancy and the lactation period of the second study infant. The average length of supplementation was about 38 months for both groups. The average consumption of the supplement by both groups of subjects was about 85 percent.

Among the 294 women enrolled, 213 completed all the requirements for the study. We consider this highly successful in terms of a longterm supplementation project.

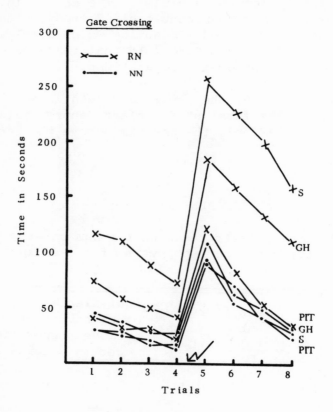

Figure 4. Effect of Growth Hormone and Pituitary Extract Administration on the Behavior of Rats in CER Tests. GH = growth hormone; PIT = crude pituitary extract; S = saliva. The arrow near the bottom of the figure (between the fourth and fifth trials) indicates when the animals received an electric shock.

Most of the subjects who left the project did so because
they moved away from the Suilin township. Very few left
the study because they could not tolerate the supplement.
The measurements made on the two groups of study infants
are:

1. Birth weight and birth length,

2. growth up to two years of age,

3. Bayley psychological measurements at eight
   months of age, and

4. nitrogen balance tests on the male study infants
   at 15 and 21 months.

As of June, 1973, we had completed the collection
of the above data. However, at the present time, only
preliminary analyses have been made of the birth weights,
birth lengths and the behavioral measurements. Four
groups of study infants were obtained and the detailed
description of each group is listed below:

1A -- first study infants of mothers taking Supplement A during lactation,

1B -- first study infants of mothers taking Supplement B during lactation,

2A -- second study infants of mothers consuming
   Supplement A during both gestation and lactation, and

2B -- second study infants of mothers consuming
   Supplement B during both gestation and lactation.

There was a decrease in the intake during the second
pregnancy in both groups of subjects. In this paper, the
birth weight data will be presented in several forms.
Because the objective of this study was to determine the
effect on the offspring of additional protein-calorie
intake during pregnancy, we used only the birth weights
of the second infants whose mothers had taken more than

50 percent of the supplement during the last five months
of the second pregnancy.

The birth weights of the first study infants are
presented in three ways: Minus the birth weights of
infants whose mothers did not take more than 50 percent
of the supplement during the last five months of her
second pregnancy and minus the subjects who dropped out
of the study and yet had provided the first study infants
(Tables IIIa and IIIb).

These data show that there was an increase in the
birth weight of the second male infants in both groups
(A and B). However, the mean increase was 162 grams for
the mothers taking Supplement A and 80 grams for the
mothers taking Supplement B. In the birth weight of
female infants, there was a slight decrease of doubtful
significance in the birth weight of the second infants
in both groups compared with the weights of the first
female infants of the same group of mothers.

TABLE III - a

Birth Weight of Boys -- Excluding the Birth
Weights of First and Second Infants of
Mothers Not Taking More Than 50 Percent of
Their Supplements During the Last Five
Months of Second Pregnancies and the
First Infants of Drop-out Cases

Study Infant	Supplement	N	Birth weight	
			Mean	s.d.
First	A	36	3103	348
First	B	39	3120	504
Second	A	41	3266	379
Second	B	47	3183	366

The second method of presentation is after deleting
the drop-out cases who provided the first study infants
(Tables IVa and IVb). In this method of analysis, the
birth weights of male and female infants from both groups
of subjects, including first infants whose mothers did
not take more than 50 percent of their supplement during
the second pregnancy are considered. The increase in the
birth weight of male infants of Group A was 172 grams
for the males and 23 grams for the females compared with
the first infants of the same groups of mothers. In
Group B, there was an increase of 91 grams in the male
infants but a decrease of 23 grams in the females.

In the third method, all the subjects were included
(Tables Va and Vb). When there were no exclusions, the
increase in the birth weight of second infants was 192
grams for the males and 32 grams for the females in
Group A, and 118 grams for the males and 5 grams for the
females in Group B.

TABLE III - b

Birth Weight of Girls -- Excluding Birth
Weights of Infants of Mothers Not Taking
More Than 50 Percent of their Supplements
During the Last Five Months of Their
Second Pregnancies and the First Infants of
Drop-out Cases

| Study Infant | Supplement | N | Birth weight | |
			Mean	s.d.
First	A	33	3139	356
First	B	42	2998	334
Second	A	40	3097	328
Second	B	42	2983	324

In all cases, whether we exclude the birth weight
of the first infants when their mothers left the study
after delivery of the first infant or we exclude the
birth weight of the first infants when their mothers did
not take more than 50 percent of the supplement, the
difference between the mean birth weights of the second
male infants is about 70 to 90 grams greater in Group A
than in Group B. This shows that protein supplementation
of the mother during pregnancy affected the birth weight
of the male infants. However, this effect did not occur
for the birth weights of female infants.

The second phase of the analysis of the birth weight
data from this study was to ask, "Does this type of supplementation decrease the incidence of low birth weight
(< 2500 grams) infants?" Table VIa presents data for male
and female infants. Before supplementation started, the
incidence of low birth weight in all the subjects was
6.7 percent. After the subjects started taking the supplement, the percentage of low birth weight infants among
the mothers who had taken Supplement A decreased to 1.2
percent, while the percentage of low birth weights in the

TABLE   IV - a

Birth Weight of Male Infants--Excluding Drop-out Cases

Study infant	Supplement	Exclusion of women taking < 50% of supplement	N	Birth weight Mean	s.d.
First	A	no	46	3094	328
First	B	no	49	3092	475
Second	A	yes	41	3266	379
Second	B	yes	47	3183	366

mothers who had taken Supplement B remained the same as before supplementation.

A corresponding analysis was made after including the birth weights of first infants of mothers who had not taken more than 50 percent of the supplementation during the last five months of the second pregnancy (Table VIb) and after including not only the mothers who did not take more than 50 percent of the supplement but also the mothers who had dropped out of the study after their first infants were born (Table VIc). Irrespective of the method of exclusion or inclusion of subjects, the results remained the same; that is, after taking Supplement A, the subjects in this group gave birth to a significantly lower percentage of infants whose birth weights were less than 2500 grams.

CHOICE OF PROTEIN SUPPLEMENT

The results of the Taiwan study, reported in the previous sections, showed that protein supplementation

TABLE IV - b

Birth Weights of Female Infants
Excluding Drop-out Cases

Study infant	Supplement	Exclusion of women taking < 50% of supplement	N	Birth weight	
				Mean	s.d.
First	A	no	47	3075	374
First	B	no	48	3006	338
Second	A	yes	40	3097	329
Second	B	yes	42	2983	324

(Supplement A) given to the pregnant women who had
marginal dietary protein intake increased the birth
weight of the infant, thus lowering the chance of having
high risk group infants (birth weight less than 2500
grams).

The protein in Supplement A was milk protein, which
is too expensive for the low socioeconomic group who
normally are in need of such supplement. If a protein
preparation could be found which was of high quality but
less expensive than milk protein, a general program of
supplementation might be possible. During the past few
years, we have been working toward such an approach.

In our rat experiments, we found that the progeny
of undernourished mothers required considerably more
food for growth and for maintenance than the progeny of
well-nourished mothers.[14] Furthermore, the progeny of
undernourished dams excreted more urinary nitrogen
although they received the same test diet as normal
progeny.[15] In 1968, we reported a nitrogen balance

TABLE V - a

Birth Weight of Male Infants

Study infant	Supplement	Exclusion of women taking < 50% of supplement	N	Birth weight Mean	s.d.
First	A	no	62	3074	315
First	B	no	63	3065	462
Second	A	yes	47	3265	379
Second	B	yes	47	3183	366

study, conducted on 11-year-old boys, using the same source of soy protein. Children from supposedly poorly nourished mothers could not maintain their body weight on the diet[16] but children from adequately nourished mothers could do so. The data from both rat and human studies indicate that food requirements, particularly proteins, can vary from one group to another.

Until recently, it appears to have been generally assumed that healthy subjects of the same age and sex and of comparable body weight and height have the same body composition and the same nutrient requirements. Our findings raise the possibility that, in humans, the protein requirements of individuals may depend upon more subtle factors than simply age, sex, body weight and height. What has been shown in rats concerning the influence of maternal diet upon the later protein requirements of the individual may be true in humans also.

Because the study of 11-year-old boys described above was done only with soy bean protein, it must be ascertained whether the same effect operates with other

TABLE V - b

Birth Weight of Female Infants

Study infant	Supplement	Exclusion of women taking < 50% of supplement	N	Birth weight Mean	s.d.
First	A	no	60	3065	364
First	B	no	62	2978	342
Second	A	yes	40	3097	329
Second	B	yes	42	2982	324

proteins. Furthermore, it must be determined whether
the requirement of amino acid supplementation with a
given vegetable protein will be the same for all individuals. To answer some of these questions, studies were
made with ADP (adequate diet progeny) and PDP (poor diet
progeny) groups using wheat as the sole source of protein,
supplemented with lysine or lysine and threonine.

In this section, data from nitrogen balance tests
on 10 to 11 year old ADP and PDP children of the same
population will be reported using wheat as the test
protein with or without amino acid(s) supplement. Information from such a study would appear useful in providing
basic data as well as practical information of value in
the eventual establishment of the proper levels of amino

TABLE VI - a

Proportion of Low Birth Weights (<2500 Grams)
Excluding the Birth Weights of First and Second
Infants of Mothers Not Taking More Than 50 Per-
cent of Their Supplements During the Last Five
Months of Second Pregnancies and the First
Infants of Drop-out Cases

Study Infants	Number of birth weights <2500 grams	Percent low birth weights
1MA + 1FA + 1MB + 1FB	10	6.7
2MA + 2FA	1	1.2
2MB + 2FB	6	6.7

The groups of study infants are identified by
number (1 or 2 for first or second), sex (M or
F) and supplementation (A or B).

acid supplementation when predominantly wheat diets are taken. At the time the present study was being planned, bulgur wheat was advocated for distribution to developing countries; for this reason, it was used in this investigation.

Ten to eleven year old boys from primary schools in Taiwan were selected as the subjects for this study. Through the excellent cooperation of the principals of several primary schools in different areas of Taipei City and its vicinity, it has been possible to select an ample number of test subjects meeting the criteria for the study and to conduct the studies during the summer vacation.

The selected boys appeared to be clinically healthy by physical examination and free from infectious and metabolic diseases before and during the study period. They were housed in a temporary metabolic ward of sufficient size to accommodate approximately 30 children at one time. Medical facilities and medical care for the

TABLE VI - b

Proportion of Low Birth Weights (< 2500 Grams) Excluding First Infants of Drop-out Cases

Study infants	Number of birth weights < 2500 grams	Percent low birth weights
1MA + 1MB + 1FA + 1FB	12	6.3
2MA + 2FA	1	1.2
2MB + 2FB	6	6.7

The identification of groups is the same as in Table VI-a.

study children were available on a 24-hour basis during
the test period. The ambient temperature in the ward
was maintained between 27° and 31°C. A playground was
adjacent to the ward. Two of the subjects' regular
teachers were hired to continue a full schedule of in-
struction to the children at the metabolic ward and to
ensure proper discipline. The teachers lived in the
ward with the subjects during the entire study period.
Thus, while the environment and activities of the subjects were carefully controlled to ensure proper and
complete collection of samples for a successful nitrogen
balance study, the strangeness of the atmosphere was
minimized and it was made as comfortable as possible.

The subjects were divided into two groups of 36
each according to the estimated animal protein intake of
their mothers during the gestation and lactation periods.
This was determined retrospectively by a food survey conducted by public health nurses from the Biochemistry
Department of the U.S. Naval Medical Research Unit No. 2,

TABLE   VI - c

Proportion of Low Birth Weights (< 2500 Grams)

Study infants	Number of birth weights < 2500 grams	Percent low birth weights
1MA + 1MB + 1FA + 1FB	17	6.9
2MA + 2FA	1	1.2
2MB + 2FB	6	6.7

The identification of groups is the same as in
Table VI-a.

Taipei, Taiwan. If the daily maternal diet had contained
meat, poultry, or fish, then the boy was classified in
the adequate diet progeny (ADP) group. If, however, the
mother had eaten the above proteins less than 10 times
per month, the boy was included in the poor diet progeny
(PDP) group.

Body weights to the nearest 10 grams were determined
on a special balance at the same time each morning and
the daily allotment of test diet was provided in proportion to that weight. The daily protein intake of each
child was 1.5 g of protein per kg of body weight and
the total daily caloric intake of each child was 85 Kcal
per kg of body weight. The actual daily food intake for
each child depended on his body weight for that day.

The basal diet offered to the subjects was in the
form of a Chinese steamed wheat bread called "mantou."
Therefore, the experimental food was not very strange to
the subjects and, consequently, there was no problem of
food rejection. Any small amount of food left by a
subject was weighed and deducted from the amount offered
to obtain an accurate measurement of the intake. The
mantou was made of Beevo (a wheat concentrate from
Pillsbury Co., Minneapolis, Minnesota, containing 22 percent wheat protein). The other constituents were:
bulgur wheat flour, 22 percent; Wesson oil, 8 percent;
sugar, 14 percent, and water, 34 percent.

The ingredients were mixed and kneaded into a bread
dough consistency, divided into loaves and steamed for
one hour. The advantages of this material for the basal
diet were that it was easy to prepare, the nutrients
were not likely to be destroyed during preparation, and
it could be weighed readily so that each individual
received an accurate portion, allocated according to his
own body weight. In addition to the above fixed amount
of protein intake from mantou for each child, an additional
amount, to make up the required total protein intake, was
provided by steamed bulgur wheat alone. Steamed bulgur
wheat was given only during lunch and dinner time. The
schedule of meals is summarized in Table VII.

Table VIII shows the protein and caloric requirements

and the sources of protein and calories in the test diets
to illustrate how the actual intake of protein and
calories was administered. Because the intake of mantou
and steamed bulgur wheat for the proper protein intake
did not provide sufficient caloric intake, the difference
was made up with a sucrose solution flavored with un-
sweetened Kool-Aid. Appropriate daily supplements of
lysine with or without threonine were administered in
the sucrose solution. The amount of sucrose in this
drink was adjusted to meet the caloric requirement set
for each individual child.

Chinese salted pickles were given during the meals
and at snack time to provide additional flavor to the
diet and to ensure the intake of additional salt. The
quantity of additional protein and calories from this
source was negligible. Each child was given one multiple
vitamin tablet daily.

Design of Experiment

This study involving 72 subjects, 36 ADP and 36 PDP
was carried out in three replicate periods. In each
period, one group of 12 ADP and one group of 12 PDP

TABLE VII

Schedule of Meals
Served During Study Period

Meal	Time	Food
		(Basal diet)
Breakfast	8 a.m.	Mantou
Lunch	12 noon	Mantou and steamed bulgur wheat
Dinner	5 p.m.	Mantou and steamed bulgur wheat
Snack	8 p.m.	Mantou

TABLE VIII

Distribution of Protein and Energy Provided in the Diet

Group	Body weight (kg)	Protein in Grams			Energy in Kcal			
		Requirement[a]	Source		Requirement[a]	Source		
			Mantou Bulgur wheat	Steamed Bulgur wheat		Mantou Bulgur wheat	Steamed Bulgur wheat	Sucrose
ADP	27	40.5	36 (89%)[b]	4.2 (10.4%)[b]	2295	1552 (68%)[c]	161 (7%)[c]	582 (25%)[c]
PDP	25	37.5	36 (97%)	1.2 (3%)	2125	1552 (73%)	47 (2%)	526 (25%)

[a] 1.5 grams protein and 85 Kcal energy/kg body weight/24 hours.

[b] Percentage of total protein intake.

[c] Percentage of total caloric intake.

children participated. The subjects and their teachers
were brought from their schools in suburban Taipei to
the metabolic ward in Taipei. The subjects were given
instructions about the housing and briefed on the daily
routine and objectives of the study, although the exact
supplement or composition of the diet to which they were
assigned was not described to them in detail. Nurses
were assigned so that each was responsible for four
children; therefore, there was sufficient around-the-clock
surveillance to ensure complete collection of urine and
stool samples and to prevent surreptitious food intake.

The subjects in each group were weighed on the morning after they had been admitted to the metabolic ward.
Each group of ADP and PDP subjects then was divided
randomly into three subgroups with each subgroup consisting of four children. The test diets received by the
subgroups were:

Subgroup I	Basal diet
Subgroup II	Basal diet + 2 grams L-lysine-HCl/child/day
Subgroup III	Basal diet + (2 grams L-lysine-HCl + 1 gram L-threonine)/child/day

A total of 12 ADP and 12 PDP subjects were in each subgroup after completion of three replications of the study.

Each experiment consisted of a three-day priming
period to accustom the subjects to the test diet and to
the environment. They were fed the test diet from the
first day of admission to the ward through the ninth day.
Nitrogen balance was calculated on specimens collected
from days four through nine. It may be questioned whether
the length of the collection period in the study was sufficient. In previously unpublished work, we found that
differences between groups with respect to nitrogen
balance data were demonstrated equally well in the period
between day four and day nine as in a longer period from
day four to day 21. Therefore, we adopted the shorter
period for the present work. A test period of similar
length was used by Barness.[17]

Urine was collected and analyzed daily for total
nitrogen. Analyses of feces were made on pooled three-
day samples. Carmine dye was given to the children in
the morning of every third day to ensure the proper
separation of the fecal samples to match their respective
three-day balance periods. Creatinine was determined to
have a baseline figure and to provide some indication
about the completeness of urine collections. Nitrogen
was analyzed in aliquots of the urine and stool samples
according to the modified method of Lang.[18]

Results

The ADP group was taller, heavier and had greater
head circumferences than the PDP group (Table IX). The
mean ages of the two groups were 10 years, 6 months for
the ADP group and 10 years, 7 months for the PDP group.
Therefore the differences in body size were not due to
differences in age. Since our test period was too brief

TABLE IX

Means and s.e. of Anthropometric Measurements of
ADP and PDP Children

Group	N	Height (cm)	Head circumference (cm)	Initial body weight (kg)
ADP	36	135.4 ± 0.6	53.1 ± 0.2	27.3 ± 0.5
PDP*	35	132.0 ± 0.8	52.0 ± 0.2	25.0 ± 0.4
		$P < 0.001$	$P < 0.001$	$P < 0.001$

*One child became ill after two days in the metabolic
ward; therefore only 35 subjects were included in
the measurements.

to allow meaningful weight changes these data are not presented.

Nitrogen balance was calculated as nitrogen intake minus urinary and fecal nitrogen (Table X). Among these children selected for the basal diet, the ADP group were in mean negative nitrogen balance of -18.2 mg of nitrogen per kg per day; this value is significantly different ($p < 0.05$) from the mean value of -32.6 mg N/kg/day for the PDP group. Among children receiving the basal diet supplemented with lysine, the ADP group was in nitrogen equilibrium. This is significantly different ($p < 0.01$) from the mean negative balance of -19.6 mg N/kg/day in the PDP group. The tests using basal diet supplemented with both lysine and threonine showed a mean positive nitrogen balance of $+5.6$ mg N/kg/day in the ADP group and a mean negative nitrogen balance of -9.3 mg N/kg/day in the PDP group; again the difference is significant ($p < 0.05$). Daily creatinine excretion was essentially the same in both ADP and PDP groups either with or without supplementation (Table XI).

TABLE   X

Nitrogen Balance mg N/kg/day in ADP and PDP
Children on Wheat Diet with or without
Amino Acid (S) Supplementation

Group	Basal diet	Basal diet + lysine	Basal diet + lysine + threonine
ADP	-18.2	0.0	$+5.6$
PDP	-32.6	-19.6^{*}	-9.3
	$P < 0.05$	$P < 0.01$	$P < 0.05$

*11 subjects in this group; each other group contained 12 subjects.

Conclusion

Despite the empirical definition of the ADP and PDP groups used in the present study and in the previous study,[16] it is interesting to note that groups defined by such criteria have differences in physical development as well as nitrogen balance. We cannot attribute such differences in the offspring to the influence of maternal diet alone because the complete dietary history of the mothers in both the ADP and PDP groups cannot be ascertained nor can we be sure of the roles played by other environmental factors. Furthermore, the dietary histories of the two groups of boys since birth must be expected to be different. Nevertheless, the present results indicate that the two groups of children have different patterns of food utilization which is in harmony with the previous results of animal studies. The findings of the present study agree also with the previous human study[16] in that both indicate that the ADP group and PDP group perform differently on the same dietary regimen.

One objective of the present study was to determine whether the nitrogen balance index of both groups of children fed wheat protein could be improved by lysine

TABLE   XI

Daily Creatinine Excretion mg N/kg/day in
ADP and PDP Children (Means and s.e.)

Group	Basal diet	Basal diet + lysine	Basal diet + lysine + threonine
ADP	19.9 ± 0.5	19.5 ± 0.7	20.3 ± 0.4
PDP	20.8 ± 0.5	20.3 ± 0.5*	20.7 ± 0.4

*The number of subjects in this group was 11;
each other group contained 12 children.

supplementation, further improved by lysine and threonine
supplementation and particularly whether the response of
the two groups to dietary supplementation was the same.
Among the ADP children, supplementation of lysine to
wheat produced significant improvement (p < 0.02) in
nitrogen balance over wheat diet alone; supplementation
of lysine and threonine to wheat showed further, but
statistically non-significant, improvement in the average
nitrogen balance indexes. Similar results were obtained
in the PDP group. The beneficial effect of joint administration of threonine and lysine to the test diet has
been observed also by Daniel.[19,20] However, like ours,
his results showed no significant difference with the
number of subjects used.

On the basis of data available in the literature,
it is apparent that further studies must be conducted to
determine whether threonine used in conjunction with
lysine will produce a significant added effect. Furthermore, although there is general agreement that supplementation with lysine alone improves the nutritive value of
wheat protein,[17,21-28] the amount of lysine to be added
to wheat may depend on the socioeconomic or dietary
histories of users of the product.

In our present study design, all the boys received
the same quantity of mantou, independent of body weight.
Additional wheat protein required to complete their
daily protein requirement was provided by steamed bulgur
wheat. Consequently, the heavier boys received a greater
proportion of their protein in the form of steamed bulgur
wheat. This design was less than optimum because of the
diminished digestibility of steamed bulgur wheat grains
compared to that of the bulgur wheat flour and the Beevo
wheat protein concentrate used in making the mantou.
Stool samples showed evidence of varying amounts of un-
digested bulgur wheat grains. Therefore there was a
tendency for the heavier boys to have less total protein
available in digested form compared with the smaller boys.

This disadvantage does not interfere with interpretation of the results except to reduce somewhat the
apparent differences between the ADP and the PDP groups.
In general, the ADP boys were heavier than the PDP boys

and, therefore, on the average, obtained a greater proportion of their protein in the form of less digestible steamed bulgur wheat. Therefore, the ADP group had a somewhat lower quantity of amino acids available for absorption than the PDP group. Despite this unintended dietary disadvantage, the ADP group exhibited better nitrogen balances than the PDP group.

From the present studies, it is obvious that the PDP group do not handle wheat protein in the same manner as the ADP children; the PDP excreted more nitrogen than the ADP. The need for a limiting amino acid like lysine may be different between these two groups so that a level of added lysine that can bring most ADP children into nitrogen equilibrium may fail to do so in PDP children. In the present study, the addition of threonine in conjunction with lysine improved the nitrogen balances of the PDP group, although they remained in negative balance, while the ADP group was brought into positive nitrogen balance. These data indicate that protein requirement, as measured by the nitrogen balance index, is different for ADP and PDP children.

However, there are a few reports[29-31] indicating that the addition of lysine does not increase nitrogen retention. The extent to which the discrepancies between these reports may be due to lack of uniformity of nutriture of test subjects or dosage of essential amino acids added and calories offered is uncertain. The experimental conditions varied from one investigator to another. However, in none of the previous studies was the possible effect of differences in protein utilization and essential amino acid requirement by the two groups of subjects with different dietary histories recognized.

An ancillary result of the present study, mentioned above, is the finding that incompletely digested grains of bulgur wheat were present in almost all stool specimens examined. Therefore, any plans to introduce steamed or boiled bulgur wheat into the dietary pattern of a population in forms resembling cooked rice, do not appear feasible. Only ground bulgur wheat has sufficient digestibility to warrant its widespread use.

The results of the present study indicate that the protein requirements of ADP and PDP subjects, within a given population, are different and, therefore, that any program involving the fortification of vegetable protein with amino acids should be preceded by tests utilizing the subjects for whom the product is intended.

REFERENCES

1.   Chow, B.F. and Lee, C.-J.: Effect of dietary restriction of pregnant rats on body weight gain of the offspring. J. Nutr., 82:10, 1964.

2.   Hsueh, A.M., Simonson, M., Kellum, M.J. and Chow, B.F.: Perinatal undernutrition and the metabolic and behavioral development of the offspring. Nutr. Rep. Int., 7:437, 1973.

3.   Simonson, M., Sherwin, R.W., Anilane, J.K., Yu, W.Y. and Chow, B.F.: Neuromotor development in progeny of underfed mother rats. J. Nutr., 98:18, 1969.

4.   Simonson, M. and Chow, B.F.: Maternal diet, growth and behavior. In: Nutrition and Intellectual Growth in Children, Bulletin 25A, Association for Childhood Education International, Washington, D.C., 1969.

5.   Lee, C.-J. and Chow, B.F.: Effect of maternal dietary restriction on the nitrogen metabolism of the offspring. Fed. Proc., 24:568, Abstract 2418, 1965.

6.   Blackwell, B.-N., Blackwell, R.Q., Yu, T.T.S., Weng, Y.-S. and Chow, B.F.: Further studies on growth and feed utilization in progeny of under-fed mother rats. J. Nutr., 97:79, 1969.

7.   Simonson, M. and Chow, B.F.: Maze studies on progeny of underfed mother rats. J. Nutr., 100:685, 1970.

8. Hanson, H.M. and Simonson, M.: Effect of fetal undernourishment on experimental anxiety. Nutr. Rep. Int., 4:307, 1971.

9. Hsueh, A.M., Simonson, M. and Chow, B.F.: The importance of the period of dietary restriction of the dam on the behavior and growth in the rat. J. Nutr., 104:37, 1974.

10. Rider, A.A. and Simonson, M.: Effect on rat offspring of a maternal diet deficient in calories but not in protein. Nutr. Rep. Int., 7:361, 1973.

11. Hsueh, A.M., Simonson, M. and Chow, B.F.: The effect of protein quality in the maternal diet on reproduction and development of the progeny. Proceedings of the IX International Congress of Nutrition, Mexico City, September, 1972 (in press).

12. Stephan, J.K., Chow, B., Frohman, L.A. and Chow, B.F.: Relationship of growth hormone to the growth retardation associated with maternal dietary restriction. J. Nutr., 101:1453, 1971.

13. Simonson, M., Hanson, H.M., Roeder, L.M. and Chow, B.F.: Effects of growth hormone and pituitary extract on behavioral abnormalities in offspring of undernourished rats. Nutr. Rep. Int., 7:321, 1973.

14. Hsueh, A.M., Blackwell, R.Q. and Chow, B.F.: Effect of maternal diet in rats on feed consumption of the offspring. J. Nutr., 100:1157, 1970.

15. Lee, C.-J. and Chow, B.F.: Protein metabolism in the offspring of underfed mother rats. J. Nutr., 87:439, 1965.

16. Chow, B.F., Blackwell, R.Q., Blackwell, B.-N., Hou, T.Y., Anilane, J.K. and Sherwin, R.W.: Maternal nutrition and metabolism of the offspring: Studies in rats and man. Am. J. Pub. Hlth., 58:668, 1968.

17. Barness, L.A., Kaye, R. and Valyasevi, A.: Lysine
     and potassium supplementation of wheat protein.
     Am. J. Clin. Nutr., 9:331, 1961.

18. Lang, C.A.: Simple microdetermination of Kjeldahl
     nitrogen in biological materials. Anal. Chem.,
     30:1692, 1958.

19. Daniel, V.A., Leela, R., Doraiswamy, T.R. et al.:
     The effect of supplementing a poor Indian ragi
     diet with L-lysine and DL-threonine on the digesti-
     bility coefficient, biological value and net
     utilization of the proteins and on nitrogen re-
     tention in children. J. Nutr. Dietit., 2:138,
     1965.

20. Daniel, V.A., Doraiswamy, T.R., Venkat Rao, S. et
     al.: The effect of supplementing a poor wheat
     diet with L-lysine and DL-threonine on the digesti-
     bility coefficient, biological value and net
     utilization of proteins and nitrogen retention
     in children. J. Nutr. Dietit., 5:134, 1968.

21. Bricker, M., Mitchell, H.H. and Kinsman, G.M.:
     The protein requirements of adult human subjects
     in terms of the protein contained in individual
     foods and food combinations. J. Nutr., 30:269,
     1945.

22. Keuther, C.A. and Myers, V.C.: The nutritive value
     of cereal proteins in human subjects. J. Nutr.,
     35:651, 1948.

23. Hoffman, W.S. and McNeil, G.C.: The enhancement
     of the nutritive value of wheat gluten by supple-
     mentation with lysine, as determined from nitrogen
     balance indices in human subjects. J. Nutr.,
     38:331, 1949.

24. Bressani, R., Wilson, D.L., Béhar, M. et al.:
     Supplementation of cereal proteins with amino acids.
     III. Effect of amino acid supplementation of wheat
     flour as measured by nitrogen retention of young
     children. J. Nutr., 70:176, 1960.

25. Bressani, R., Wilson, D., Béhar, M., Chung, M. and
    Scrimshaw, N.S.: Supplementation of cereal pro-
    teins with amino acids. IV. Lysine supplementa-
    tion of wheat flour fed to young children at dif-
    ferent levels of protein intake in the presence
    and absence of other amino acids. J. Nutr., 79:
    333, 1963.

26. Graham, G.G., Placko, R.P., Acevedo, G., Morales, E.
    and Cordano, A.: Lysine enrichment of wheat
    flour: Evaluation in infants. Am. J. Clin.
    Nutr., 22:1459, 1969.

27. Rice, H.L., Shuman, A.C., Matthias, R.H. and Flodin,
    N.W.: Nitrogen balance responses of young men
    to lysine supplementation of bread. J. Nutr.,
    100:847, 1970.

28. Graham, G.G., Morales, E., Cordano, A., and Placko,
    R.P.: Lysine enrichment of wheat flour: Prolonged
    feeding of infants. Am. J. Clin. Nutr., 24:200,
    1971.

29. King, K.W., Sebrell, W.H., Severinghaus, E.L. and
    Storvick, W.O.: Lysine fortification of wheat
    bread fed to Haitian school children. Am. J.
    Clin. Nutr., 12:36, 1963.

30. Pereira, S.M., Begum, A., Jesudian, G. and
    Sundararaj, R.: Lysine-supplemented wheat and
    growth of preschool children. Am. J. Clin.
    Nutr., 22:606, 1969.

31. Reddy, V.: Lysine supplementation of wheat and
    nitrogen retention in children. Am. J. Clin.
    Nutr., 24:1246, 1971.

SOME STATISTICAL CONSIDERATIONS ON THE USE OF

ANTHROPOMETRY TO ASSESS NUTRITIONAL STATUS

Harvey Goldstein

From the National Children's Bureau
London, England

Although nutritional differences between individuals
or populations may lead to anthropometric differences,
the existence of anthropometric differences between individuals or populations does not necessarily imply the
existence of nutritional differences. One cannot, for
example, in a nutritional supplementation program, use
anthropometric status in a simple minded way as a criterion of nutritional success. Rather, the question that
must be asked is how well, if at all, we can use anthropometry to predict nutritional status. What is required
is some function, or functions, of anthropometric and
other relevant measurements that predict either specific
or general nutritional states.

We must know the population to which any such function applies before it can be used in practice. The
basic units of a population may be defined as either
individual people or, for example, villages or even
countries or races. In the remainder of this paper,
references will be made only to units consisting of
individual people, although many of the remarks apply
also to units that are aggregates of individuals. However, there is no reason to believe that we will obtain
similar functions for individuals and aggregates of individuals, nor that the predictions obtained will be
equally useful. It may happen, for example, that a

prediction may be poor on an individual level, but may
give useful results when applied to a well defined
group of individuals. For instance, an alteration in
food supply may have affected the physical growth of
some children in some groups of a population. A comparison, say, of the average values of appropriate
anthropometric measurements may enable us to identify
those groups that have suffered or benefited from the
alteration. On the other hand, these same measurements
may be so insensitive that, when taken on a single child,
they are of little value in an individual clinical
assessment.

Having established such a function for a particular
population, it is important to study how invariant it is,
not only over time in the same population, but also over
different populations. This implies that studies should
be designed to be replicated over several populations
and several points in time. It should be emphasized
that it is important not only to make predictions but
also to have a theoretical scientific basis for understanding how and why such predictions work. Without
such scientific support these predictions can only be
supported by what Popper[1] refers to as an 'ad hoc' explanation. In the long run, success of the public
health concern with prediction will depend on success
of the scientific search for explanation.

The above remarks are rather general. Attention
will now be directed to two more detailed questions:
sampling and the presentation of results in terms of
population standards.

SAMPLING

To obtain useful predictions of nutritional status
from anthropometric variables implies that we need adequate samples of all the anthropometric categories (age
groups, racial groups, etc.) for which nutrition will be
measured. The problems of obtaining adequate samples
are particularly acute at some ages and in certain types
of populations. In rural populations, for example,
there may exist no adequate sampling frame and

considerable resources may be necessary to prepare one.
For children of school age, an adequate sampling frame
may often be available in the form of school attendance
records, but even where schooling is universal, the
sampling of preschool children can pose difficult problems even in highly industrialized populations.
Clearly, there is little point in having sophisticated
techniques for measurement and analysis if the wrong
individuals are being studied. Often this is the weak-
est point of epidemiological studies of this kind.

Within a basic sampling frame, two methods of
studying individuals can be distinguished: the cross-
sectional method where each individual is measured once
only and the longitudinal method where each individual
is measured on more than one occasion. Where some individuals are measured once only and some more than once,
this is known as a mixed longitudinal study. The choice
between these different methods will be determined
partly by constraints of time, etc., and partly by the
need to obtain particular types of information. If,
for example, we are interested in the relationship between nutritional states at different ages, this implies
a longitudinal design. If we require information about
rates of change with age ('velocities'), then again a
longitudinal design would usually be chosen. Where any
of the three designs will give appropriate estimates,
the choice depends on relative costs, although it may
not always be easy to quantify all the costs involved.
A discussion of some of these problems is given by
Patterson.[2]

POPULATION STANDARDS

The most common use of population standards is to
assess clinically those individuals who fall outside
'acceptable' limits. Once such individuals have been
identified they may either be treated or followed to
obtain more information, or both. If we could measure
nutritional status directly, the problem would be to
define such limits for the nutritional constituents of
diet, separately and in combination, for a given

population. Where, however, only indirect measures of
nutrition are available, standards based on these should
in some way be related to the limits that ideally would
be based on direct nutritional measurements.

   This is indeed how many of the ordinary standards,
such as those for height and weight, are used. It is
not, in this case, the fact of, say, short stature that
interests us, but, for example, what short stature represents in the way of undernutrition or specific
hormonal lack. This immediately raises the problem
that a single anthropometric measurement may be acting
as a proxy for any one of a number of different etiologies that could influence it, and different 'limits'
will be the most sensitive for different etiologies.
This is especially true when combinations of measurements such as height and weight are being considered,
where particular combinations of categories will be
associated with different normal or abnormal etiologies.
For example, a child may be very heavy for his age because he has relatively too much fat for his size or
possibly because he is simply a large size, or for
other reasons; and different reasons have different
implications for health and treatment.

   Also we should understand that to choose a particular function of two or more measurements, solely
because it possesses certain mathematical properties
(such as being uncorrelated with another measurement),
is no guarantee that the chosen function is the best
one for the purpose. It has been shown, for example,
that three different weight for height classifications
(percentile position, relative weight, and a ratio index)
produce very dissimilar classifications of 'overweight'.[3]
Of 354 children classified as 'overweight' by at least
one of these classifications, only 43 percent were so
classified by all three, and 29 percent by only one.
Without an external criterion, such as the prediction
of a specific nutritional disease state, there can be
no satisfactory basis for preferring any one function
to any other.

In different populations for which different standards are applicable, the limits used may differ also. In a well nourished population, for example, those individuals above, say, the 98th percentile of weight for height may be suspected of obesity, whereas in a badly nourished population interest might center on individuals below the 10th percentile, because they are likely to be suffering from malnutrition. Great care has been taken when attempting to apply standards derived from one population or group to another.

Sometimes the application of standards derived from economically privileged groups to less privileged groups is justified on the grounds that, given genetic similarities between groups, it may be assumed the less priviledged individuals are not achieving their potential and the privileged standards represent this potential. The argument runs that the privileged standards are appropriate for the underprivileged group. No doubt if the underprivileged individuals had spent their lives in a privileged environment, they might have achieved the standards appropriate to the privileged group, although this could be disputed. The point, however, is that they have not lived in such an environment, and there is no reason to expect that two individuals with the same measurements from such different environments will both enjoy the same state of health, or require the same treatment. It is quite possible, for example, that a child who would be regarded as abnormally small in a privileged group, is actually well adapted to an underprivileged environment, where according to standards appropriate to that environment he is not abnormally small. It may indeed be harmful to attempt to 'treat' him by increasing his nutritional intake. This could cause a maladaptation to the environment in which he remains. These suggestions are, of course, only possibilities but, until much more is known about such problems and about whether genetic homogeneity can be assumed, it would appear that where standards are applied to a group, they should be the ones appropriate to that group, i.e., derived from individuals belonging to that group. When combinations of separate measurements are involved, it may well be the case that different combinations are appropriate for different groups, and the above remarks become even more pertinent.

Broadly speaking, population standards may be used in two different ways. In one way, they may be regarded as a 'screen' by which individuals can be assessed readily and selected for further scrutiny if they fall outside predetermined limits. The decision about where to put the limits will depend usually on a number of complex considerations. Not only will it be necessary to know which particular ranges of nutritional states are implied by particular anthropometric measurements, but this knowledge will have to be related to the resources available for following or testing selected individuals. It could be both inefficient and unethical to select either too few or too many individuals in relation to the resources available. To satisfy such considerations, it is clear that the information provided by the population standards must be appropriate to the group concerned.

The second use of population standards is to assess an entire group of individuals, for example, a geographic or social unit. Nearly always, there will be significant anthropometric differences between such groups. As with individuals, usually the case for 'treating' an underprivileged group, for example, by a nutritional supplementation program, will rest on a number of considerations. In many circumstances, the information supplied by anthropometry alone may be insufficient to enable a satisfactory decision. It may be necessary to collect more detailed information or possibly to conduct experimental programs. It seems worthwhile to emphasize that anthropometric differences per se are not necessarily bad. By themselves, such differences are not usually a sufficient justification for a nutritional program; 'bigger' does not necessarily mean 'better'.

Also we are faced with the problem of how many subpopulations or groups should be recognized as worthy of separate standards, and how far group differences

should be ignored in providing overall standards cover-
ing groups or subpopulations. The answer to this question depends partly on convenience and cost. The
convenience lies in the ease of use of the standards
and the cost in the sampling and measuring of individuals from different subpopulations. Also it will
depend on the size and stability of the differences
between the groups.

Little attention seems to have been paid to this
latter point and there appear to be certain inconsistencies in the ways that standards have been defined
in particular populations. Child growth standards, for
example, are usually presented separately for the two
sexes but not separately according to the number of
siblings in a child's family although at many ages this
variable accounts for bigger differences than does sex.

Figure 1 presents standards based on a sample of
seven year old children from the National Child Development Study[4] classified by parity (number of previous
births to the child's mother) and social class (Regis-
ter General's classification, United Kingdom). For
children in social class I or II with no older siblings, the third percentile is equivalent to about the
seventh percentile for the total population and the
population third centile is equivalent to about the
first centile for this group.

Whether the increased precision of classification
is considered worth the effort of preparing and using
separate standards for subpopulations will depend on
how much better such standards are as 'screening'
instruments. That is, how much more sensitive and
selective they are in identifying those individuals who
have, or will develop, the nutritional state that the
particular anthropometric function is designed to detect. In some cases, it may be possible to quantify
the costs and benefits involved.

Of course if standards are to have practical use,
they need to be presented in as simple a way as possible and therefore without too many categorizations.
Also there is a need to make categorizations that are

easily measurable. There have been recent attempts to
extend birthweight standards by including more categorizations, for example, the birthweight standards of
Tanner et al.[5] are categorized by gestation, mother's
weight, and mother's height.

    In addition, we need to search for new and conve-
nient methods of presenting standards with many categorizations. The maximum number of ways of categorizations that seems to have been attempted so far is four;
this could probably be increased to five or six, using

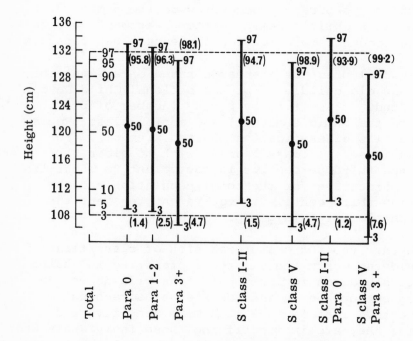

Figure 1. Percentile standards from The National Child
Development Study for height at seven years for all
United Kingdom children. The figures in brackets give
percentile values for subpopulations equivalent to the
3rd and 97th percentiles in the total population. "Para"
refers to parity and "S class" to social class.

the methods adopted by Tanner et al.[5] in the birthweight standards, and in the standards for children's
height allowing for the height of their parents.[6]

Most standards are still concerned with a single
anthropometric measurement. Bivariate standards have
been produced but in terms of presentation the second
measurement effectively reduces by one the number of
categorizations possible and, hence, there is a limit
to the adequate presentation of true multivariate standards in graphical form. There is no limit to the use
of multivariate standards (assuming that they can be
estimated) if they are used by applying the appropriate
calculations, rather than looking up positions on a
chart. However, the usefulness of bivariate or multivariate standards is not clear.

Usually bivariate standards have been presented in
terms of probability ellipses, where each ellipse contains 97 percent, 90 percent, etc., of the population.
However, unlike univariate standards, it will not, in
general, be true that we may expect a one-to-one increasing or decreasing relationship between the anthropometric measurement and the nutritional state to which
it is related. For example, all those individuals
lying on the 97th percentile ellipse will not, in general, have the same risk of a particular nutritional
state. Instead, we can draw 'contours' representing
what this risk is estimated to be, for all points on
the bivariate diagram. Thus, for nutritional purposes
these contours will be more relevant than the general
bivariate standards. Such contour diagrams (including
multivariate ones) provide a general method of predicting nutritional state, and single indices or functions
of measurements may be seen as attempts to summarize
such diagrams. From a theoretical point of view,
whereas these functions may be less sensitive than the
most general kind of prediction, considerations of data
collection, presentation, and use may often favor the
simpler approach in a practical situation.

REFERENCES

1.   Popper, K.R.: Conjectural knowledge: my solution
     of the problem of induction. In: Objective
     Knowledge, an Evolutionary Approach. London,
     Oxford University Press, 1973.

2.   Patterson, H.D.: Sampling on successive occasions
     with partial replacement of units. J. Roy.
     Statist. Soc. B, 12:241, 1950.

3.   Newens, E.M. and Goldstein, H.: Height, weight
     and the assessment of obesity in children.
     Brit. J. Prev. & Soc. Med., 26:33, 1972.

4.   Davie, R., Butler, N.R. and Goldstein, H.: From
     Birth to Seven. London, Longman, 1972.

5   Tanner, J.M. and Thomson, A.M.: Standards for
     birthweight at gestation periods from 32 to 42
     weeks, allowing for maternal height and weight.
     Arch. Dis. Childh., 45:566, 1970.

6.   Tanner, J.M., Goldstein, H. and Whitehouse, R.H.:
     Standards for children's height at ages 2-9
     years, allowing for height of parents. Arch.
     Dis. Childh., 45:755, 1970.

BIOLOGICAL REFERENCE SYSTEMS IN THE ASSESSMENT OF NUTRITIONAL STATUS

Napoleon Wolański

From The Polish Academy of Sciences, Warsaw, Poland

The assessment of nutritional status in the laboratory or clinic has a different significance and is based on methods other than those used in the assessment of nutritional status in epidemiological investigations. In this paper, we shall pay attention to the second type of assessment.

PREMISES FOR ASSESSING NUTRITIONAL STATUS

As we shall show below, the assessment of nutritional status of either an individual or a whole population cannot be arbitrary. In other words, there are not, nor can there be, universal measures of nutritional status, because all the properties of an individual, or of a population, that indicate nutritional status result from the past development of every individual and generation and, to some extent, the development of preceding generations. This occurs because of metabolic influences from the mother on the development of the fetus and also on the fetal gametes, which is a kind of "heritability" on the mother's line. In addition, the sensitivity of the body to environmental factors is determined genetically as well as being influenced by maternal effects on the metabolism of the fetus. Ontogenetic development, per se, is a process of adaptation of the body to the external environment. The

chance to survive, to bear, and to rear offspring allows
the prolongation of the species. The process of adaptation to the environment is not separate for isolated
factors, but it is integrated with the necessity to maintain homeostasis of the organism. Consequently, the body
is somewhat "comprehensive" in its response to a single
stimulus, or to multiple stimuli.

The conviction that the response of the organism
to a set of modifying factors is complicated probably
arose as a result of the necessity to investigate many
environmental factors. The conviction that the extent
of the modifications in the organism is proportional to
the variable factors in the environment is not, however,
self-evident. To some extent, these modifications depend
on the sensitivity and resistance of the organism to
environmental stimuli, on the stage of development and
on the past experience of the organism in respect of the
particular stimuli (Figure 1).

Nutrition is only one of many environmental factors.
It is impossible to understand its effects and to evaluate
the nutritional status of the organism without considering physical activity and other aspects of the mode of
life, climatic conditions (particularly temperature),
and other factors. The same level of nutrient intake can
produce more than one "nutritional status" depending on
the whole ecological situation of the individual. Consequently, traits should be chosen for assessing nutritional
status that have a self-evident interpretation. Furthermore, it is necessary to select a standard population,
as a biological reference system, that will have a known
relationship to the population to be assessed.

To demonstrate a few aspects of this complicated
problem, some actual situations will be described. They
will illustrate the essence of the problem better than
theoretical reasoning.

Example 1. In villages of poor but relatively
homogeneous nutrition, the similarity between children
and their parents is much greater than in a town with
a higher living standard and a large assortment of foodstuffs[1] (Table I). We excluded illegitimate children,

who are most common in towns, by serological identification of families. The existence of a similar phenomenon
was confirmed later by others.[2,3] This phenomenon is
probably closely related to the mode of nutrition.

The following hypotheses may be formulated:

(i)   Under urban conditions, or in social strata
      with a higher standard of living, there are
      greater differences between children and
      their parents in living conditions and the

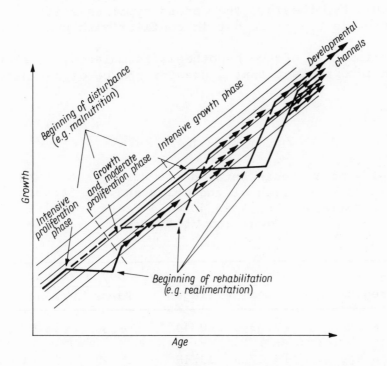

Figure 1. The degree to which developmental disturbances
are compensated according to the time of the disturbance,
e.g., malnutrition, and its duration before rehabilitation. The figure shows the theoretical possibilities of
returning to a single developmental channel; all these
possibilities occurred in the author's data.[4]

mode of nutrition than in less affluent social strata.

(ii) In towns and under better living conditions, a greater amount and diversity of foodstuffs is eaten. Consequently, there are more alternative metabolic pathways even if there are no genetic differences between the groups.

Let us consider the first hypothesis. If, in better off strata, the children were, for example, more accelerated in their growth in stature, they would be shifted in the correlation field and, in principle, this should not significantly alter the correlation coefficients (Figure 2a). The analysis of recorded data has enabled us to reject this hypothesis. The second hypothesis intuitively seems probable and it is not in conflict with known facts.

Irrespective of the hypothesis formulated, it should be taken into account that a greater chance of deviations

TABLE I

Results of χ^2 tests between the Parent's Stature and Age at the Birth of the Child and the Stature of the Children When Aged 4 to 14 Years in Polish Towns and Villages[5]

Data from parents	Girl's stature		Boy's stature	
	Towns	Villages	Towns	Villages
Mother's age	18.59	40.44**	6.50	14.75
Father's age	15.87	23.48	18.29	20.41
Mother's stature	29.13*	48.73**	26.76*	28.94*
Father's stature	13.63	17.83	20.44	25.14

*P <0.05; **P <0.01.

will exist under urban conditions, because of potentially
diverse pathways of development. Therefore, the assessment of an urban population will be much more complex
from the viewpoint of nutritional status, than that of a
rural population. Correspondingly, the assessment is
more complex for individuals of social strata with better
living conditions than of individuals living under poorer
conditions. When there is undernutrition, deficiencies
in stature, body weight, adipose tissue, hemoglobin
concentration or iron level in the blood will provide

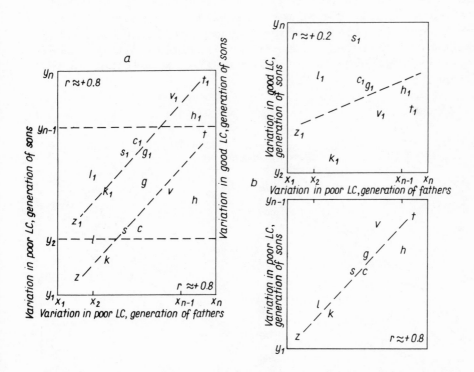

Figure 2. When living conditions are modified between
generations, the correlation coefficient between conditions, across generations, may be the same, if there is
a random shift of the subjects towards greater values of
the trait investigated (Example a). The correlation
coefficient will be altered if there is a change in
position within the group, due to different pathways of
development (Example b). LC = living conditions.

more certain proof of poor nutritional status, despite
interrelationships with infection, than under conditions
in which, on an average, there is no deficiency of food
in the population. In the latter situation, multidirec-
tional deviations in development may occur, in at least
the above-mentioned traits.

Example 2. In a population with a relatively high
standard of living, children of parents with discordant
stature categories are taller than those whose parents
are of medium stature, although both groups of children
are living under similar conditions (Figure 3). This
effect is clear in older girls. It may be assumed that
heterozygotic girls are more sensitive to advantageous
living conditions, including better nutrition. Similar
results for children in exogenous families (large matrimonial radius) have been reported from several investi-
gations.6-9

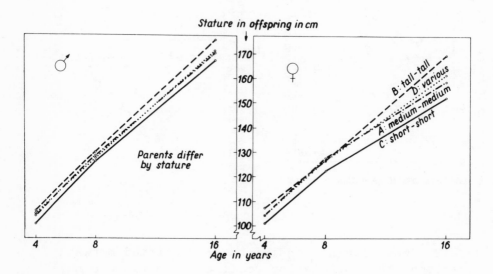

Figure 3. Stature in boys and girls aged 4, 8 and 16
years in urban (Szczecin) families in which the parents
differ according to stature categories. Symbols: A =
medium-medium (both parents of medium stature); B = tall-
tall; C = short-short; D = various.

In undernourished populations with a low standard of living, it was found, however, that children of parents with discordant stature categories ("various" in Figure 4) are shorter than those whose parents are of medium stature. Apparently, heterozygotic children are more sensitive to food deficiency, expressed in a reduction of stature, than children who are homozygotic for a given trait.

Two important conclusions, relevant to the problems of assessing nutritional status, can be made:

(i)   Heterozygotic organisms are more prone to the influences of environmental factors, including nutrition.

(ii) In populations with a greater tendency to cross-breeding or outcrossing, individuals who are heterozygotic for the trait under investigation will show greater nutritional effects. The greater the proportion of such individuals in

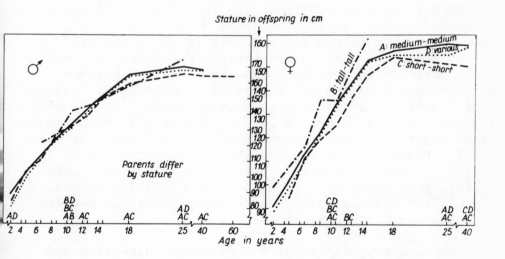

Figure 4.   Stature in offspring aged 1.5 to 70 years in rural families in which the parents differ in stature grouping. The symbols are the same as in Figure 3. The letters show significant differences.

the population, the greater the effect of
nutrition on the population. This is true for
undernutrition, excess nutrition and incorrect
or unbalanced nutrition.

In rural populations, and in poorer social strata,
there is less mobility. This is shown by data relating
to migration and matrimonial radius. Consequently, there
is a greater chance of homozygosity. The traits of these
individuals will be deflected less by the influence of
ecological conditions than in urban dwellers, where the
opposite conditions occur. This is why heterozygous
individuals, and populations heterogenous for the given
trait, react more strongly than homozygous populations
to a stimulus, e.g., nutritional intake, that is the same
in type, intensity and duration. Thus the given level of
traits, on the basis of which nutritional status has been
assessed, may reflect not only the strength and duration
of the stimulus, but also sensitivity to the stimulus.

This phenomenon is, however, complicated by another
factor: the percentage of the given feature developed
during intrauterine life proportional to the magnitude
of this feature in the adult. Intrauterine development
of the feature occurs without the fetus having direct
contact with the environment external to the mother's
body. Features that develop mainly in the postnatal
period are more sensitive to environmental factors than
those that develop mainly during fetal life. Features
exhibiting a high degree of postnatal development include
body weight (95 percent after birth), stature (71 percent),
shoulder width (70 percent), recumbent length (65 percent),
nose height (62 percent) and chest circumference (61 percent). Features that develop largely in the prenatal
period include: minimum frontal width (45 percent after
birth), face width (43 percent), head width (40 percent),
head length (39 percent), head circumference (36 percent)
and nose width (36 percent).

In heterozygotic subjects, with good nutrition and
appropriative living conditions, the features of the first
group (those that grow mainly after birth) increase,
whereas the other features (those that grow mainly before
birth) diminish (Figure 5). This complicates analyses

of the effects of nutrition on anthropometric variables.

Example 3. The traits used to measure biological
status, including nutritional status, depend also on
genetic polymorphism and the influence of sociological
factors.[10] On the other hand, among various countries,
there is a relationship between the statures of women
and per capita annual gross national product value
(Figure 6). On the other hand, the differences between
mean statures at the same mean income level without doubt
express genetic differences in addition to differences
in the standard of living.

It is extremely important that the reference population for the assessment of nutritional status be selected
so that any comparisons made will reflect nutritional
factors, not genetic ones.

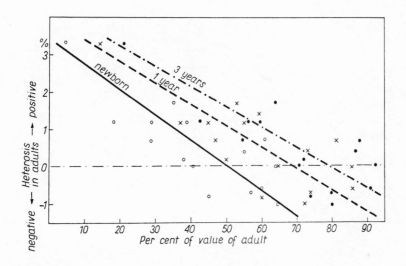

Figure 5. The percentage by which a given feature is
greater (positive heterosis) or smaller (negative heterosis) in groups of subjects from exogenous and endogenous
families in relation to the degree of development of the
trait during fetal life.[11]

Example 4. Sensitivity to the influence of ecological factors differs with the stage of development, being
greater in periods during which development is rapid.[12,13]

A number of sets of data support this claim. Analyses
similar to that shown in Figure 6 have demonstrated that
sensitivity to the basic components of the diet is, on
the population scale, particularly great during the first
two to four years after birth and during the pubertal
period (Figures 7 and 8). The curves of these relationships differ across age and across groups of nutrients.
Animal protein intake shows a continual effect on stature
and the Quetelet index. However, total protein intake
and the caloric value of the food have optimal values
above which stature and the Quetelet index decrease in
all or in some age classes (Figures 7 and 8). This is
probably less due to negative effects of surplus protein

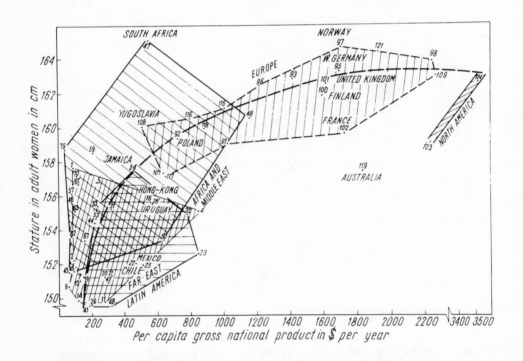

Figure 6. Per capita annual gross national product and
stature in adult women in various countries.[10]

or calories and more to the mode of nutrition associated
with economic status. Food of high caloric value is
common in countries with a low economic standard.   In
many developed countries there is a widespread desire
for a slim figure, and moderate eating is recommended
for health reasons.

These findings lead to two important conclusions:

(i)   Not all stages in human development are equiva-
      lent for the assessment of nutritional status.

(ii) A uniform level of development of a particular
      trait is not necessarily evidence of the same
      degree of nutritional status among individuals
      in terms of the nutritional needs of the body,
      even when there is no reason to believe that
      the individuals differ genetically.

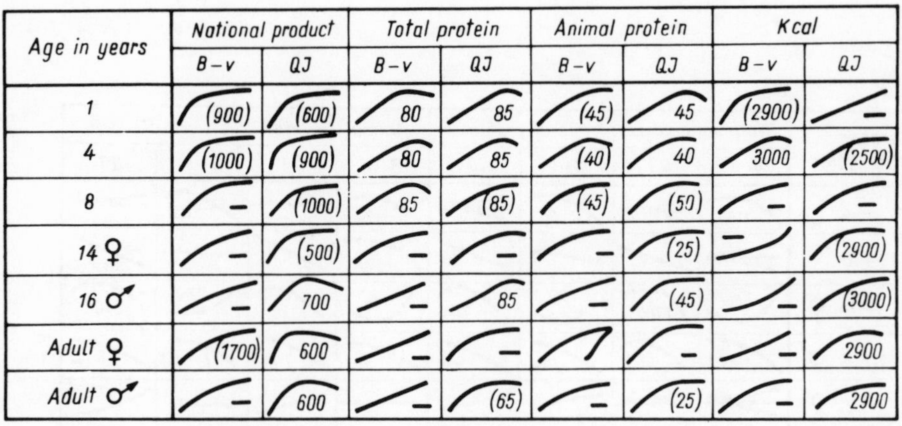

Age in years	National product		Total protein		Animal protein		Kcal	
	B−v	QJ	B−v	QJ	B−v	QJ	B−v	QJ
1	(900)	(600)	80	85	(45)	45	(2900)	—
4	(1000)	(900)	80	85	(40)	40	3000	(2500)
8	—	(1000)	85	(85)	(45)	(50)	—	—
14 ♀	—	(500)	—	—	—	(25)	—	(2900)
16 ♂	—	700	—	85	—	(45)	—	(3000)
Adult ♀	(1700)	600	—	—	—	—	—	2900
Adult ♂	—	600	—	(65)	—	(25)	—	2900

Figure 7.   Shapes of smoothed curves of stature (B-v) and
Quetelet index (QJ) in relation to the per capita annual
gross national product, the total protein and animal
protein supply and total calories in various countries.[10]
The Quetelet index is weight (g)/stature (cm).   The
numbers indicate the critical points at which the curves
change in direction.   Those in brackets indicate values
after which stature and the Quetelet index do not increase.

Normative data must be effective when employed to identify nutritional deficiencies, or disturbances, or disease. Particular attention must be given to conditions in which considerable nutritional deficits occur, or in which the basic food requirements are met, but problems arise due to the deviations from developmental pathways. Often an abnormal nutritional status has to be identified under rapidly changing conditions. Checking, by anthropometric methods, will often be too late, especially in regions being rapidly industrialized, or areas where agriculture becomes intense during a brief period. The first signs of these changes may be unexpected in their directions and combinations.

Example 4. In assessing the biological or nutritional status of the population one should consider various overlapping effects. These effects might be interpreted erronously as changes due entirely to a particular eating pattern.

The population of the city of Szczecin provides such

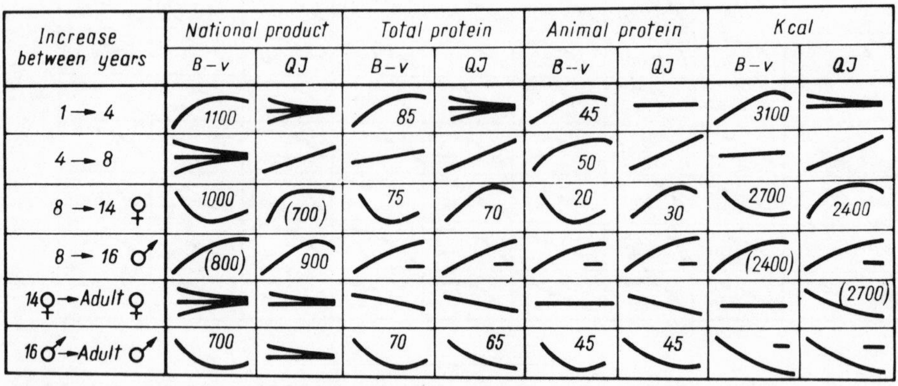

Figure 8. Shapes of smoothed curves of increments in stature (B-v) and Quetelet index (Q^J) versus per capita annual gross national product, the total protein and animal protein, and total calories in various countries.[10] The numbers have corresponding meanings to those in Figure 6.

an example. Between 1945 and 1947, this city became
Polish and 97 percent of the population was changed as
a result of this alteration in its political control.
This city, that had already been inhabited, was settled
again by people coming from various parts of Poland.

Initially, from the biological viewpoint, they were
not a firmly knit population. That is, they were not a
group of families living in one area for at least a generation and interbreeding relatively freely. In this
period, there was a very high mortality compared with
other Polish cities and districts, both in Szczecin and
in the district of Szczecin (Figure 9). The children of
Szczecin, however, were above average in stature and
other indices for the children, that might be regarded
as measures of biological or nutritional status, were
satisfactory also. At that time, the living conditions
of the population were poor due to lengthy hostilities.

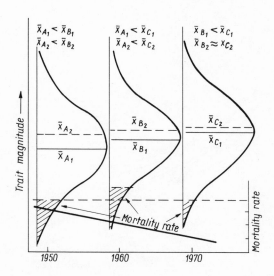

Figure 9. A schematic representation of changes in
infant mortality related to the distribution of a continuous trait. This effect can mask the real influence
of nutrition on the distribution of values of the trait.

It was deduced that these differences were due to
a high degree of heterozygosity in the population; the
matrimonial radius was 270 km.[14] Also, it is possible
that smaller infants had died and that those who survived
had high developmental indices.[15] As soon as heterozygosity was reduced, the infant mortality index dropped
from 120 to less than 30 (Figure 10). The indices of
biological status of the children were reduced also.
Presumably, this occurred because this generation was
born from parents born and living in Szczecin. Consequently, the matrimonial radius was small and, presumably, there was less heterozygosity. This situation is
represented by the model in Figure 9. These findings
show that the assessment of nutritional status on the
basis of features such as stature or body weight can be
misleading. In the first ten years, when the nutritional
status of this population was poor, the commonly used
indices were increased. During the second decade, when

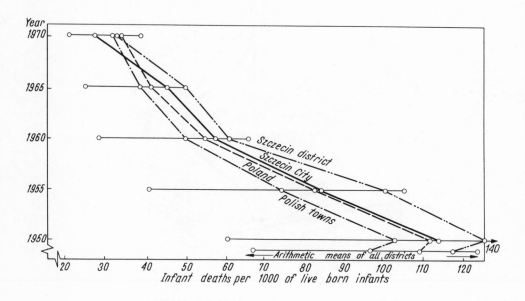

Figure 10. Infant mortality indices in the city and
district of Szczecin, for Poland as a whole, and for
Polish towns from 1950 to 1970.

nutrition was adequate, there was no statistically significant increment in the indices of nutritional status.

Summation

Two extremely important problems should be stressed:
the strength of the environmental stimulus and the degree
of sensitivity of the organism. These factors are presented in a schematic fashion in Figure 11. The degree
of reception and the changes elicited may vary for two
identical stimuli. On the other hand, stimuli differing

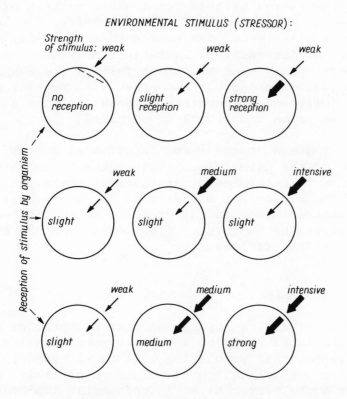

Figure 11. A schematic representation of the strength
of the stimulus, its reception by the organism and the
response of the organism.

in intensity may induce the same effects in the organism.
Finally, the extent of the changes in the organism may be
proportional to the intensity of the environmental
stimuli. The ratio between the nutritional status of the
organism and the food intake is not necessarily fixed.

It is necessary to investigate environmental factors
precisely because the responses and state of the organism
must be related to these as well as to food intake. In
analyzing environmental effects, we must take into account
not only food intake, but also the entire complex of
factors (for example, sanitary levels, temperature, motor
activity) that may overlap factors relating to food intake. Unfortunately, these factors are usually ignored
in nutritional investigations. Nevertheless, it is well
known that the energy balance for a given pattern of food
intake depends on the level of motor activity, the level
of basal metabolism under the common thermal conditions,
the numbers and intensity of bacterial invasions, among
other factors. Only when all these factors are examined
can the ecological phenomena concerned with nutrition be
understood fully and an accurate diagnosis of the nutritional status of an individual be obtained.

Hence, instead of the term "nutritional status," I
suggest use of the term "biological status." An attempt
could be made to determine by statistical means the proportion of the biological status that is due to nutritional
factors. What is measured, at present, as nutritional
status reflects the pattern of food intake and the effects
of non-nutritional factors.

THE SELECTION OF NORMATIVE DATA

The selection of a population that can provide normative data should be based on the above considerations.
For convenience, this population will be referred to as
the "standard population." The selection of traits and
age classes to be measured is of fundamental importance
because it affects the applicability of "normal values."

The concept of normality in a biological context,
refers to the normality of the relationship between an

organism and the environment in which a gene pool, specific for a given population or individual, has been to
some degree stabilized. It includes the concept of good
health or effective psychophysical condition, taking into
account the genotype and the medical and nutritional
development of the population or individual. Normality
cannot be defined adequately by the use of negative con-
trasts as in: "normality is that which is neither pathological nor abnormal." Such negations contribute nothing.

In this formulation, the biological meaning of
normality can, to some extent, be identified with the
concept of prevailing conditions and the statistical
average. The following facts provide evidence that this
occurs. Mutations occur and, for the population as a
whole, they are random in direction. The influence of
environmental factors exhibits bipolarity, at least, for
example, across ranges of temperature and food intake.
Consequently, they act on the organism in various directions. Extreme external environmental conditions cause
a spreading of the central tendency which is identified
with the average external environment. The normative
data form an empirical standard.

A conservative formulation of the norm is insufficient, because factors in the environment, in particular
nutritional deficiency, do not permit use of the complete
genetic potential. When there is a positive social tendency to improve the biological status of a population by
individual and population hygiene, normative data should
not be equal to the actual statistical average.

ECOLOGICAL FORMS OF THE GENE POOL

From the above remarks, it can be concluded that
there are at least three concepts of the norm:

(i)   Genetic-individual. A set of genes conditions
        the normal development of an individual. This
        set does not contain detrimental recessive
        genes. Detrimental dominant genes are usually
        lethal.

     (ii)   Ecological-individual. This is a phenotypic
                description of an individual, developing in
                a given country and time, under and in equi-
                librium with optimal environmental conditions.
                The criteria of normality of parents and of
                intrauterine environment should be added to
                this concept.[2,16]

     (iii)   Genetic-population. There is a gene pool
                that is specific for a particular population.
                This concept relates the norm to traits common
                to a group of individuals.

Among these three concepts, only the latter can be
applied to man, in whom many experiments are unacceptable.
With skillful procedures, this third formulation can include elements of the preceding ones.

Summing up: In the formulation of a norm for an
individual, a particular person should be evaluated by
comparison with normative data. The normative data would
be derived from "standard" individuals who are normal in
respect of suitable criteria. The genotypes of the
assessed subject, and those of the individuals on whom
the standards are based, should not differ significantly.
The standard should be derived from a panmitotic population that is genetically balanced. Although, from the
genetic point of view, this population should be in a
balanced polymorphic state and not under heavy selective
pressure, from the demographic point of view this population may be increasing in numbers. This would be a
measure of its biological effectiveness. Moreover, the
standard population should be one that, from the subjective viewpoint of contemporary science, is enjoying
optimal environmental conditions.

Many phenotypes can be present in a population with
the same gene pool. However, this is true only for non-
panmitotic populations that are not in polymorphic balance. This is most clear in populations where there are
differential fertility rates and, therefore, positive
selection of individuals. This is responsible for changes
in gene frequencies. It is permissible to conclude that
in an urbanized society with limited differences in

climate, nutrition and physical activity, and with
restricted migration, a given gene pool would give rise
to one phenotype. In this situation, the population
could provide a definite standard, and deviations from
the standard in individuals of the same population could
be interpreted as measures of the influence of external
environmental factors.

This procedure is the more correct because it is
not based only on recording the biological state of the
individual or the population being assessed. An assessment would be made of the advantage or disadvantage of
the individual's environmental conditions. Stated another
way, one would be investigating the biological reaction
of the organism to the environment.

From the above it can be concluded that, from the
medical viewpoint, it is insufficient to describe physical conditions of individuals or populations without reference to the circumstances under which these conditions
have been attained. When external factors are ignored,
assessment of biological conditions cannot lead to effective preventative or therapeutic actions.

Due to the above factors, norms must be formulated
in a manner that enables investigations of the dynamics
of development. Initially the state of development of a
child was referred to the calendar age, later to various
developmental ages.[17] The next step is to consider the
discordance between developmental ages and calendar age.
This results from differences between children in their
rates of development. Assessment of the dynamics of
development, in which the magnitude of the changes is
analyzed as a function of time, is one modern way to
investigate development.[18]

GENERAL PRINCIPLES FOR THE DESIGN OF NORMATIVE DATA

A definite reference system is the basis of any
assessment method. Such a biological reference system
must be based on appropriately collected and suitably
analyzed data. In general terms, the normal range must
be based on a healthy population, living under good

conditions that prevail in the region. A population with
high morbidity or nutritional deficiency should not be
selected as a standard or reference population.

It is easy to justify the necessity to establish
"local norms." Certainly, there is more than one type
of development.[19-26] At present we cannot separate these
types a priori. Perhaps any separation is only partially
possible due to recombination of genes and environmental
variations. Therefore, adoption of local norms should
be considered a first approximation. They reflect the
biological adaptation of the population to the environment. A single composite set of normative data for all
towns of a country, for instance, is biologically
unacceptable.

One of three possibilities has to be chosen:

(i)   To assume there are no fundamental differences
       between the gene pools of the various towns
       and to choose as a source of normative data
       the town with the best environmental condi-
       tions.

(ii)   After appropriate investigations, it may be
       concluded that some towns or areas have dif-
       ferent gene frequencies. If so, separate
       sets of normative data have to be prepared
       for them.

(iii) Finally, separate norms can be supplied for
       each population.

Although each of the above procedures may prove correct, usually it will be incorrect to supply one set of
normative data for all towns of a given country. This
will not be a single population because there will not
be free random exchange of genes among the towns. The
populations of all or some of the towns will not form a
panmitotic population that is balanced genetically. This
is one case when the notion of a statistical population
does not correspond to that of a biological population.

Finally, there may be a disproportion between the

biological condition of a rural child and an urban child.
In most countries the adoption of the same norms for rural
and urban children cannot be recommended, even if these
areas do not differ in gene frequency. In Poland, if
urban norms were used in the assessment of rural children,
80 percent of the latter would appear retarded or malnourished. Conversely, if rural norms were applied to
urban children, nearly all the latter would appear to be
developing satisfactorily. The adoption of an average
norm would not be correct either. Such a choice would
only change the percentages. It would be known in advance
that about 70 percent of the rural children would be
below the norm and about 70 percent of the urban children
would be above the norm.

PURPOSE AND APPLICABILITY OF NORMATIVE DATA

Normative data for the growth and development of
children and youth aim at control of the biological condition of the population and its individual members.
However, a norm should not be a target of action.

Normative data can be established on the basis of
investigations of a well nourished population with respect
to quantity, composition and frequency of meals.[27,28]
If the opposite were the case, the subjects could be malnourished. In the latter case, the development indices
would be too low (Figure 6). If the needs of a child for
protein or other dietary components were established with
reference to these indices, this would encourage the per-
sistence or even the appearance of "malnutrition."

Normative data must assist the attainment of the
aims of the public health service. From the point of
view of such a service, "target" norms, i.e., oversize
normative data should be suggested. Such norms would
warn the physician of slight states of arrested development and would assist prophylaxis and the improvement of
population health.

The above remarks concerning somatic features require
qualification, because it is assumed that a taller child,
or one who matures earlier, is "better." This conclusion

is by no means self-evident. In preindustrial periods
muscular strength was important. At present, however,
with industrialization and automation, intelligence is
more highly valued than muscular strength. However,
there is a rational basis for considering a large or
early maturing child to be biologically "better." When
living conditions are advantageous and less arduous,
children grow taller and mature earlier than their coevals
developing under poorer living conditions (Figure 5).
However, it has not been shown that tall or early maturing children are more useful or of greater value for the
development of cultural or social needs.

Of course, not all problems are as simple. For
example, the vital capacity of the lungs and hemoglobin
concentration are interrelated.[29] Therefore, normative
data cannot be supplied for the hematocrit index without
taking into account vital capacity, ventilation, cardiac
output and climate. All these factors influence the
supply of oxygen to tissues. Therefore, we cannot analyze
the function of any one system because compensatory feed-
backs occur among systems (Figure 12). This is a typical
instance of the need to apply environmental standards,
and to perform complex investigations. There are additional difficulties. Thus, norms for the above variables
for children of an industrial city cannot be derived from
the same children because the supply of oxygen to the
tissues is abnormal, due to industrial pollutants (Figure
13). These normative data have to be established on an
urban population with a similar gene frequency living in
a similar climate and altitude, but without air pollution.
In other words, we have to select a population, as the
basis for normative data, that has not deviated from
natural development due to detrimental factors.

Finally, one may ask, "What should be used as normative data if, in differing environments, early development and adult traits are similar, but the course of
development differs?" Flattening of the foot, measured
by the angle of talus inclination, is an example (Figure
14). Which values should be adopted as the standard?
In this decision, the cause of the marked differences
observed should be considered. Growth changes in the
arch of the foot are related to locomotion, the amount

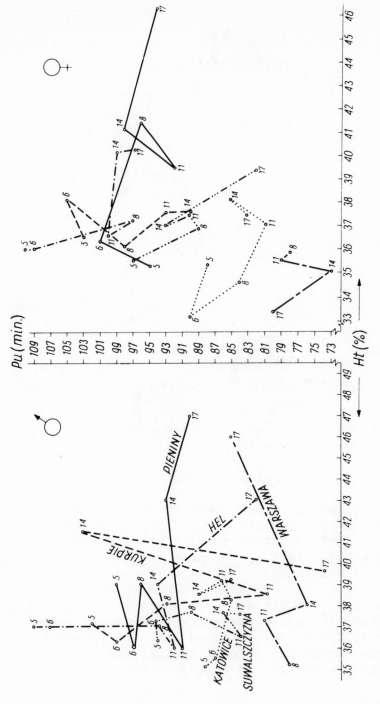

Figure 12. Data showing the interrelationships between pulse rate (Pu/min) and hematocrit (Ht) in children aged 5 to 17 years. 29 The hematocrit is expressed as a percent of normal. Katowice is a large, heavily polluted industrial city.

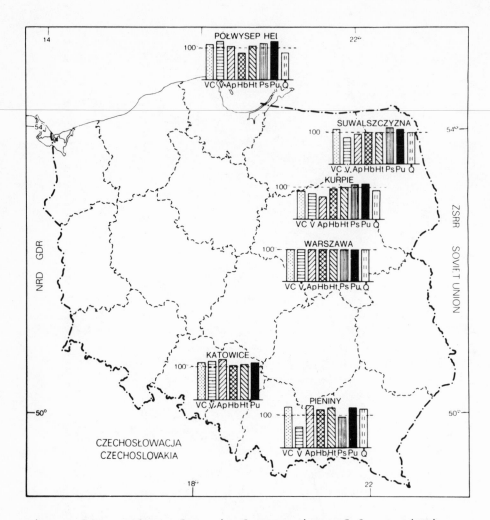

Figure 13. Values for vital capacity of lungs (VC),
ventilation (V̊), duration of apnoea (Ap), hemoglobin
concentration (Hb), hematocrit index (Ht), stroke
volume of the heart (SV), pulse rate (Pu), and the
cardiac volume heart/minute (Q̊) in children and youth
aged 5 to 17 years from various regions of Poland.
Comparisons have been based on percentages of means from
Warsaw children.[29] Hel = seaside area, Pieniny = low
mountain area, Katowice = polluted industrial center,
Suwalszczyzna and Kurpie = rural agricultural areas,
and Warsaw is the capital of Poland.

of walking and whether shoes are worn. It is important for the adult to have a properly arched foot. However, it is difficult to decide which should be considered normal values.

The assessment of obesity is difficult also. In principle, a large body weight in proportion to stature does not necessarily indicate obesity. A slim leptosomic or ectomorphic individual cannot have a large body weight, in proportion to stature, even with considerable development of subcutaneous fat. On the other hand, a mesomorphic person with an athletic body build may exhibit a large body weight, in proportion to stature, despite scanty subcutaneous fat. One of the best ways to assess this phenomenon is to apply three centile grids: for stature, body weight, and subcutaneous fat thickness. If the fat thickness centile corresponds to the stature centile for the individual and if it does not exceed that for his body weight, one cannot conclude that he is obese. When body weight has a higher centile than stature, this

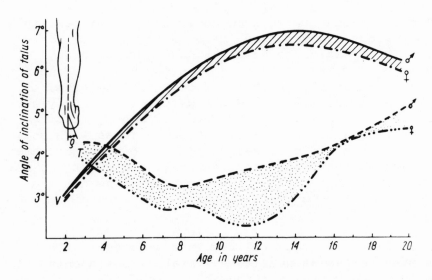

Figure 14. Changes in the angle of inclination of the talus /∢S / of children aged 2 to 20 years in rural (V) and urban (T) environments.[30]

indicates a strong body build that could be caused by
muscular development. Only when the centile for body
weight considerably exceeds that for stature, in the
same individual, and the fat thickness centile corresponds
to that of body weight may one speak of obesity.

KINDS OF ASSESSED PROPERTIES AND AGE GROUPS

A paramount problem is to choose properties that are
essential for the assessment of nutritional status. There
are two stages in this procedure:

(i)   Preliminary control to determine whether the
      individual is in good physical condition.
      Normative data play a selective role.

(ii) Finding the causes of the deviation and its
      degree. Standards have a discriminative sig-
      nificance in this application. There is a need
      for norms concerning, for example, non-specific
      immunity and the efficiency of various organs,
      such as the kidneys.

The relation between some physiological properties,
nutrition needs and lean body mass is important. Technically, it is easier to consider body surface area,
because other traits, for example, hemoglobin and the
hematocrit, expressed in terms of body surface area,
demonstrate environmental and sex differences. Body surface area is a significant indicator of energy requirements. But some of the relationships decrease with age,
for example, the ratio of hemoglobin to body surface area.
This indicates that, with age, the organism functions more
economically. This occurs in various environments and
this comparatively constant relation can be used to assess
health.[29]

Analyzing, in a general way, the anthropometric
traits used to assess nutritional status, one should
recommend primary traits rather than those that represent
several properties. It is best to avoid measurements
such as biacromial diameter or arm circumference, which
include osseous, muscle and fat tissue. Each of these

tissues has its specific development dynamics (Figure 15).
When such "collective" dimensions are considered, one
does not know which tissue is responsible for the changes.

The problem of selecting age classes so that the data
will be most effective for prevention is a separate, complex problem. Some ages can be grouped. In other developmental periods, annual increments should be considered.
In addition, in many populations it is difficult to establish the correct ages of children. That is why, in principle, the following age classes are recommended:

-- neonate.

-- the time at which the child has four deciduous
   teeth erupted. The average age is close to one
   year.

-- from when the child has four deciduous teeth
   erupted until eight are erupted. This is from
   about one to two years.

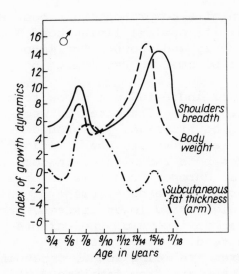

Figure 15. Developmental changes in shoulder breadth
(biacromial diameter), body weight and subcutaneous
fat.[31,32]

-- from the end of the preceding period until four
   permanent teeth are erupted (about two to seven
   years).

-- the third stage of development of secondary sex
   characters. An increase in body fat has occurred
   but the adolescent spurt in stature has not
   begun.

-- the adolescent spurt in stature.

   With this system, the age tolerance is unimportant
but comparisons between and within populations are possible.

THE DESIGN OF NORMATIVE DATA

   These standards should be regional in character and
they should exceed the current central tendency in the
population. Such standards may be termed "target" normative data. The construction of normative data of this
type requires the following steps:

1. Selection of a comprehensively best environment
   that exhibits optimal living conditions for the
   given country and epoch. The inhabitants of
   this area become the "standard" population.

   (i) It was indicated above how to proceed when
   there are large differences between areas, e.g.,
   towns and countryside. It should be stressed
   that, in recent years, the division into social
   strata has been reevaluated in some countries.
   In Central European countries, for instance, the
   differences in the development of children
   between higher and lower classes, based on the
   profession or income of the parents, are decreasing. Instead, differences in the development
   of children are being found, depending upon
   whether they are from families with several
   children or one child. There are differences
   also in the physical development of children
   depending on whether they attend daily or

weekly crèches, or are educated at home, even
if other socioeconomic circumstances do not
differ between the groups.[33-35]

(ii) The population must come from a healthy
environment, e.g., low pollution.

(iii) In principle, it must represent one
ethnic group.

(iv) The population must be biologically and
genetically in equilibrium without high fertility or mortality rates and with low migration
indices. However, it must not be a small
isolate.

(v) Often, other considerations have to be
taken into account. If, for example, normative
data are needed for blood pressure, a population
must not be chosen for which the living conditions favor hypertension. If normative data
concerning endurance fitness are required, they
should not be derived from a population with
limited motor activity.

2. A random group of the inhabitants should be
included in the sample. Where there are day
or preschool centers, sometimes investigations
are limited to the children attending these.
However, these children are usually less well
developed because they live under poorer living
conditions.[34,36] Standards based on these
children would be too low. To a lesser extent,
the same is true for preschool children. In
the selection of urban children aged over 14
years, we may find children who come from the
neighboring rural area each day and those who
live in boarding schools. Children who come from
the rural area are usually less well developed
than the general population (Figure 16).

3. A representative sample is selected at random
among the inhabitants according to a system by
which each individual has an equal chance of

entering the sample based on the structure of
the population relative to age and sex. In
principle, the sample should include at least
four percent of the population. The final sample
must include at least 30 subjects of each group
based on age, sex and social stratum. It is
usual to select a random sample that includes
30 percent more persons than the minimum. Ten
percent of them may not attend the examinations
and 20 percent are needed for the selection discussed below (point 5).

4.   The examinations should be made only in the
      morning hours because of diurnal variations.
      Appropriate and frequently calibrated instru-
      ments and universally accepted methods should
      be employed. It is best to make the examinations

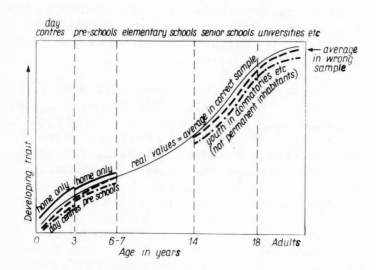

Figure 16. A schematic representation of the values
for a trait in a properly selected sample and the
directions of differences in subgroups. Selection based
on institutions, e.g., day care centers or high school
student dormitories, can lead to erroneous sets of
normative data.

during seasons when feeding deficiencies do not occur. In many countries, early autumn is most appropriate.

5. It is necessary to reject those persons examined who are abnormal. Consequently, there must be interviews and additional examinations of the persons selected. This procedure increases the average values in the standard group compared with the general population. In constructing standard values, centiles should be used because some traits have skewed distributions. It is frequently desirable to use the following centiles: 5th, 15th, 35th, 50th, 65th, 85th and 95th. Subjects outside the range from the 5th to the 95th centiles need special examinations. The ranges from the 35th to the 65th and from the 15th to the 85th assist the assessment of deficiencies and surpluses, e.g., weight. Also it is appropriate to refer stature to the midparent value.

Even if the above restrictions are followed, there are qualifications to the applicability of the normative data:

(i)   the data cannot be allowed to become obsolete due to secular trends,

(ii)   there should be no significant differences in gene frequencies, nor marked differences in environmental conditions or in socioeconomic conditions between the standard population and the population under examination, and

(iii)   the measurements should be made according to the same methods and using the same instruments as those used to obtain the normative data. The instruments must be checked frequently.

ANALYSIS OF CHANGES IN AN INDIVIDUAL

OR A POPULATION TO ASSESS NUTRITIONAL STATUS

Even the best normative data are insufficient, if appropriate methods are not used to assess the growth and physique of an individual or a population. At present, longitudinal observations are made more commonly and these data allow better assessments of biological status.

An important method of analysis is by growth rate indices which may be presented in various ways according to their purpose. To compare the rates of development of children, or of various features in the same child, the size of a feature can be expressed as a percentage of the initial magnitude, e.g., at birth or in the first month or first year (GRI_I in Figure 17). Another method is to express the size of the feature at a certain age as a percentage of the final adult size (GRI_F). Alternatively one can express the increment in a given period, e.g., an annual increment, as a percentage of the initial magnitude (GRI_L). A further possibility is to express an increment as a percentage of the total increase in the feature from birth to maturity (GRI_D). Each method has its advantages and disadvantages.[32,37,38]

Recently, an age-independent method has been developed to assess the development of subcutaneous fat in relation to body weight. This method is of significance for assessing states on the borderline of average. It is based on centiles for subcutaneous fat and body weight established on the same population. The relation between the centiles for subcutaneous fat thickness and for body weight are used to recognize obesity (Figure 18).

In making an assessment, one plots the centile for subcutaneous fat thickness on the y-axis. It is best to use the sum of the thicknesses of 3 to 10 skinfolds measured in various places of the body[39] and centiles for body weight. The weight centile is plotted on the x-axis. The position of the plotted point determines whether the individual is considered average, or stocky, etc. This method needs to be tested on other populations and age

classes, outside the range 2 to 27 years, to define its
usefulness. Particular care is required to interpret
the extreme zones (below the 20th and above the 80th
centiles) because these zones are not regularly curved
(Figure 18). This method is designed to assess individuals, and successive points for an individual can be
plotted. This method may be used also for rapid assessment of a group or population.

It was the necessity to analyze changes occurring
in the arithmetic mean, or median, for the population,
that suggested a new method of analysis. This consists
in the calculation of correlation coefficients between
the positions within the group occupied by the subjects
at the first and second examinations (Figure 19).

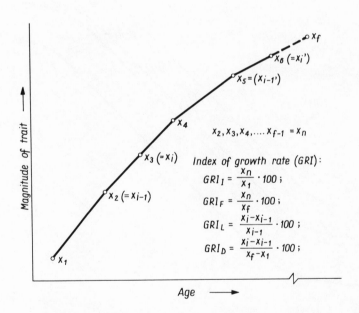

Figure 17. Growth rate indices (GRI). GRI_I relates
present size to initial size. GRI_F relates present
size to final size. GRI_L relates an increment to the
size at the beginning of the interval for which the
increment is calculated. GRI_D relates an increment to
the total increase from birth to maturity.

The investigation of changes of indices of biological or nutritional status of a population, and the correlations between the positions of subjects between successive examinations, offer opportunities for assessing intra-populational phenomena.

The assessment of the nutritional status of an individual and, more so, of a population is complex. However, when the mechanisms responsible for the effects of environmental factors are known, it will be possible to assess exactly the nutritional component in biological status.

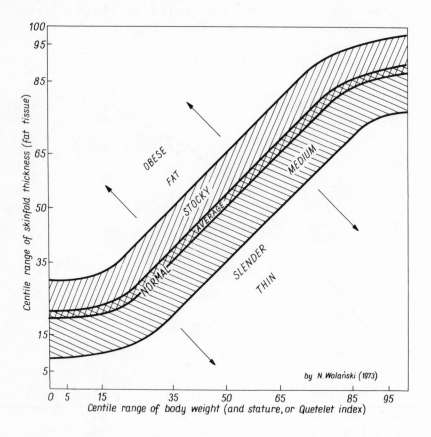

Figure 18. A diagram showing a method of assessing obesity or malnutrition using centiles for subcutaneous fat thickness and body weight.

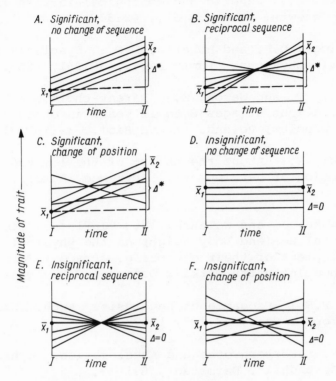

CHANGES OF TRAITS IN A POPULATION:

A. Significant,
   no change of sequence

B. Significant,
   reciprocal sequence

C. Significant,
   change of position

D. Insignificant,
   no change of sequence

E. Insignificant,
   reciprocal sequence

F. Insignificant,
   change of position

Magnitude of trait

Figure 19. Schematic changes of nine features in a
population, during a certain period, and changes in the
position of various persons within the population.
Statistically significant changes of the arithmetic
mean occurred in schemes A, B and C. There are no
changes in the arithmetic mean in schemes D, E and F.
In schemes A and D, some individuals maintained their
position in the group. This is shown by a statistically
significant positive correlation coefficient between
the values at time I and time II. In schemes B and E,
there were marked changes in position in the population.
This is expressed in a statistically significant negative
correlation coefficient. In schemes C and F, there was
a chaotic shifting of position from time I to time II.
Consequently, the correlation coefficient was close to
zero.

REFERENCES

1. Wolański, N.: Genetic and ecological factors in
    human growth. Human Biol., 42:349, 1970.

2. Bainbridge, D.R. and Roberts, D.F.: Dysplasia in
    Nilotic physique. Human Biol., 38:251, 1966.

3. Bielicki, T. and Welon, Z.: Parent-child height
    correlations at ages 8 to 12 years in children
    from Wrocław, Poland. Human Biol., 38:167, 1966.

4. Wolański, N.: About the theory of the limited
    direction of development. Acta Med. Auxol.,
    3:201, 1971.

5. Charzewska, J. and Wolański, N.: Influence of
    parental age and body height on the physical
    development of their offspring. Prace i Materiały
    Naukowe Instytutu Matki i Dziecka, 3:9, 1964.

6. Hulse, F.S.: Exogamie et heterosis. Arch. Suisse
    Anthropol. Gen., 22:103, 1957.

7. Nold, F.: Körpergrösse und Akzeleration. Wehrdienst
    und Gesundheit, Darmstadt, 8:1963.

8. Ferák, V., Lichardová, Z. and Bojnová, V.: Endogamy,
    exogamy, and stature. Eugen. Quart., 15:273, 1968.

9. Schreider, E.: Inbreeding, biological and mental
    variations in France. Amer. J. Phys. Anthrop.,
    30:215, 1969.

10. Wolański, N. and Stejgwiłło-Laudańska, B.: Growth
    and physique of the Polish population as compared
    with populations of other countries. Stud. Human
    Ecol., 1:58, 1973.

11. Wolański, N.: The problem of heterosis in man.
    In: Bernhard and Kandler (Eds) Bevölkerungs-
    biologie, 16. Stuttgart, Fischer Verlag, 1974.

12.   Wolański, N. and Żekoński, Z.: Social life and
      living conditions as factors in human development,
      p. 354. In: Wolański (Ed) Factors in Human
      Development. An Introduction to Human Ecology.
      Warszawa, Państwowe Wydawnictwo Naukowe, 1972.

13.   Wolański, N.: Human developmental stages connected
      with sensitivity to developmental factors, p. 483.
      In: Wolański (Ed) Factors in Human Development.
      An Introduction to Human Ecology. Warszawa,
      Państwowe Wydawnictwo Naukowe, 1972.

14.   Wolański, N., Jarosz, E. and Pyżuk, M.: Heterosis
      in man. Growth in offspring and distance between
      parents' birthplaces. Soc. Biol., 17:1, 1970.

15.   Wolański, N.: Biological status and integrative
      processes of population of town Szczecin in 25
      years of Poland. Przeglad Zachodniopomorski,
      4:5, 1970.

16.   Ounsted, M. and Ounsted, C.: Maternal regulation
      of intra-uterine growth. Nature, 212:995, 1966.

17.   Crampton, C.W.: Physiological age, a fundamental
      principle. Amer. Phys. Educ. Rev., 13:3, 1908.

18.   Wolański, N.: Basic problems in physical develop-
      ment in man in relation to the evaluation of
      development of children and youth. Curr. Anthrop.,
      8:35, 1967.

19.   Meredith, H.V.: The rhythm of physical growth.
      A study of eighteen anthropometric measurements
      on Iowa City white males ranging in age between
      birth and eighteen years. Univ. Iowa Stud. Child
      Welfare, 11, No. 3, 1935.

20.   Jasicki, B.: Der Entwicklungsprocess der männlichen
      Schuljugend in Krakau. Prace i Materiały
      Antropologiczne Polskiej Akademii Umiejetności,
      Kraków, PAU, 1:1938.

21. Jasicki, B.: Further studies on school children
     developmental dynamics. Prace i Materiały
     Anthropologiczne Polskiej Akademii Umiejetności,
     Kraków, PAU, 2:1948.

22. Bayley, N.: Individual patterns of development.
     Child Develop., 27:45, 1956.

23. Reed, R.B. and Stuart, H.C.: Patterns of growth
     in height and weight from birth to eighteen years
     of age. Pediatrics, 24:904, 1959.

24. Tanner, J.M.: Growth. In: Physical Medicine in
     Paediatrics. Altrincham, Butterworths, 1965.

25. Lasota, A.: An attempt of individual analysis of
     developmental dynamics in boys since birth up
     to 3 years of age, living in Warsaw. Prace i
     Materiały Naukowe Instytutu Matki i Dziecka,
     3:173, 1964.

26. Garn, S.M. and Rohmann, C.G.: Interaction of
     nutrition and genetics in the timing of growth
     and development. Pediat. Clin. North Amer.,
     13:353, 1966.

27. Brožek, J. and Wolański, N.: Age changes and sex
     differences in body compositions in childhood and
     in youth. Wychowanie Fizyczne i Sport, 10:23,
     1966.

28. Fabry, P.: Feeding Pattern and Nutritional Adapta-
     tions. Prague, Academia, 1969.

29. Pyżuk, M. and Wolański, N.: Respiratory and
     Cardiovascular Systems of Children under Various
     Environmental Conditions. Warszawa, Państwowe
     Wydawnictwo Naukowe, 1972.

30. Wolański, N.: Environmental modification of human
     form and function. Ann. N.Y. Acad. Sci., 134:
     826, 1966.

31. Wolański, N.: Kinetics and Dynamics of Growth and
    Differentiation in Body Proportions in Children
    and Young People from Warsaw (Aged from 3 to 20
    years inclusively). Warszawa, Państwowy Zakład
    Wydawnictw Lekarskich, 1962.

32. Wolański, N.: Methods of Control of Physical
    Development of Children and Youth. Warszawa,
    Państwowy Zakład Wydawnictw Lekarskich, 1965.

33. Winter, K.: Gesundheitliche Entwicklung von
    Jugendlichen in der DDR. Berlin, Akademie Verlag,
    1962.

34. Wolański, N., Charzewska, J., Charzewski, J.,
    Kowalska, I., Lasota, A. and Miesowicz, I.:
    Physical development of family children and of
    nursery children. In: Care of Children in Day
    Centres. WHO Public Health Pap., 24:118, 1964.

35. Sälzler, A.: Ursachen und Erscheinungsformen der
    Akzeleration. Berlin, Volk und Gesundheit Verlag,
    1968.

36. Buterlewicz, H., Chrzanowska, D. and Przetacznikowa,
    M.: Psychological development of family children
    and nursery children. In: Care of Children in
    Day Centres. WHO Public Health Pap., 24:126, 1964.

37. Wolański, N.: The Decourt-Doumic morphogram
    diagnostic method as modified by the author
    applicable to children and youth from 3 to 20 years
    of age. Polish Endocrinol., 13:78, 1962.

38. Wolański, N.: Methods of Control and Developmental
    Norms of Children and Youth. Warszawa, Państwowy
    Zakłady Wydawnictw Lekarskich, 1974.

39. Pařísková, J.: Rozvoj Aktivni Hmoty a Tuku u Deti
    a Mladeže. Praha, Statni Zdravotnicke Naklada-
    telstvi, 1962.

SELECTION OF THE MINIMUM ANTHROPOMETRIC CHARACTERISTICS

TO ASSESS NUTRITIONAL STATUS*

R. Buzina and K. Uemura

From the Institute of Public Health
Rockefellerova, Zagreb, Yugoslavia

The usefulness of anthropometric measurements in
the assessment of nutritional status, and the relatively
simple techniques involved, has led to their broad
application in nutritional surveys. The information
obtained from anthropometry is considered to be useful
for two purposes:

(a)   assessment of the existing nutritional status
       by measuring total body mass and body composi-
       tion, and

(b)   the follow-up for the community of trends of
       ecological factors, including nutrition,
       through periodic measurement of anthropometric
       characteristics of growth and development.

As anthropometry becomes an integral part of nutrition survey techniques, it is necessary to select and
standardize the most useful and practical measurements.
This need is felt particularly in epidemiological surveys
when relatively large numbers of subjects have to be
examined within limited periods. Therefore a study was

* Supported, in part, by a grant from the World Health
  Organization.

271

planned with the following aims:

1.   To select the minimum anthropometric measure-
     ments to be included in epidemiologic surveys
     that are useful to assess current nutritional
     status and ecological or nutritional trends in
     the community;

2.   To study the changes in anthropometric charac-
     teristics with age to obtain provisional refer-
     ence standards for the evaluation of increments
     of growth and development in the community with
     changing ecological factors, and

3.   To standardize the methodology of the recommended
     anthropometric measurements, primarily those
     that can be made by less specialized personnel.

                              Methods

     The following anthropometric measurements were taken:
body height (stature), sitting height, body weight, circumferences (arm, chest, abdomen, calf), fatfolds (triceps, subscapular, abdomen) and biacromial and bicristal
diameters. In children less than three years old, head
circumference was measured also. The following indices
were calculated from the measured values: relative body
weight (percent of standard weight), the ratio weight/
height^2, laterality-linearity index (ratio of the sum of
the biacromial and bicristal diameters to total body
length), the ratio sitting/standing height and the ratio
biacromial/bicristal diameter. The latter was interpreted
as an index of relative masculinity-femininity.

     The study was made in a genetically homogeneous popu-
lation, one group of whom lived in a rural area and another
resided in nearby towns and industrial centers where they
had adopted an urban way of life for two or three generations. A third population group, differing in ethnic
background and known to be among the tallest in the
country, was studied for the purpose of comparison.

RESULTS

The age-sex composition of the three population groups examined is shown in Table I. For the present analysis, the results from the three groups have been pooled.

The determination of nutritional status was based on body weight after the differences in the length of the skeleton were eliminated by calculating the relative body weight using Jelliffe's reference standards.[1] The often recommended weight/height[2] index was not used because it is age-dependent after the age of three years. To compare the importance of different indices in the assessment of nutritional status (as indicated by percent of standard weight for height), correlation coefficients were calculated between each index and the relative body weight for males and females in the age groups one to three, four to six, seven to ten, and 11 to 14 years. The data are summarized in Table II.

Arm circumference, a combined measure of muscle and subcutaneous fat, shows the highest correlation with relative body weight. The abdominal circumference also shows a fairly high correlation with relative body weight.

Table I

Age-Sex Composition
of the Examined
Population Groups

Age (years)	Male	Female
1- 3	835	864
4- 6	623	583
7-10	900	843
11-14	884	764
Total	3242	3054

TABLE II

Correlation Coefficients between Relative Body Weight and Other Variables

Independent variables	Age 1-3 yrs.		Age 4-6 yrs.		Age 7-10 yrs.		Age 11-14 yrs.	
	Male	Female	Male	Female	Male	Female	Male	Female
Stature	.015	.138	-.134	.140	-.022	-.093	-.115	-.034
Sitting height	.182	.232	-.026	.274	.100	.053	.058	.126
Arm circumference	.629	.680	.335	.689	.548	.484	.588	.595
Chest circumference	.437	.547	.277	.620	.451	.342	.437	.507
Abdomen circumference	.535	.608	.344	.723	.524	.435	.522	.551
Calf circumference	.554	.609	.273	.611	.386	.327	.447	.480
Biacromial diameter	.222	.322	.060	.329	.214	.065	.150	.218
Bicristal diameter	.265	.350	.139	.346	.178	.103	.134	.170
Triceps fatfold	.414	.467	.191	.538	.427	.428	.549	.596
Subscapular fatfold	.475	.517	.236.	.574	.428	.460	.548	.582
Suprailiac fatfold	.398	.453	.259	.557	.449	.450	.550	.566
Arm muscle circumference	.483	.525	.280	.566	.467	.377	.415	.430

The correlation coefficient for the arm muscle circumference (an index calculated by subtracting the thickness of the subcutaneous fat from the arm circumference) was slightly lower than that for the arm circumference itself. The other two circumferences, viz. those of the chest and the calf, and the subcutaneous fat thicknesses, showed only moderate correlations with relative body weight. The biacromial and bicristal diameters showed comparatively low correlations with relative body weight while stature and sitting height showed the lowest correlations. The latter was expected because the effect of stature had been eliminated in the calculation of relative body weight and sitting height is highly correlated with stature.

In the next step, an attempt was made to determine whether the deletion of eight variables relating to muscle and fat thickness had a significant effect on predictions of relative body weight (the dependent variable). To obtain these results, all the 12 variables were considered first, with respect to the dependent variable. Next, the regressions of the four variables relating to "skeletal size" were considered separately. The effect of the "muscle and fat" variables on the regression was calculated by subtracting the contribution of the skeletal size variables from the contribution of all 12 variables. In each case, the "muscle and fat" variables, as a group, contributed significantly to the decrease in the standard error of the estimate of relative body weight.

To select the single measurements of "skeleton size" and of the "muscle and fat" that contributed most to the estimation of relative body weight, a stepwise regression analysis was made in which the skeleton size variables were forced into consideration first (Table III). Apparently, bicristal diameter and stature, in the age groups 1 through 6 years, and the biacromial diameter and stature in the age group 7 through 14 years are the skeletal size variables that contribute most to the estimation of relative body weights.

Among the seven variables relating to "muscle and fat," the arm circumference and abdomen circumference,

TABLE   III

The Selection of Muscle and Fat Tissue Variables
(5-9) Contributing Mostly to Regression After
Variables of Skeleton Size (1-4) Have Been
Considered First. Results from Stepwise
Regression Analyses in Which the Dependent
Variable is Relative Body Weight

Males Age 1-3 years; N = 835

Step	Variable added	R	Standard error of estimate
0	----	----	8.829
1	Bicristal diameter	.265	8.519
2	Stature	.425	8.002
3	Sitting height	.563	7.312
4	Biacromial diameter	.575	7.239
5	Arm circumference	.758	5.774
6	Abdomen circumference	.811	5.189
7	Calf circumference	.833	4.899
8	Chest circumference	.847	4.712
9	Subscapular fatfold	.854	4.623

Females Age 1-3 years; N = 864

Step	Variable added	R	Standard error of estimate
0	----	----	7.966
1	Bicristal diameter	.350	7.467
2	Stature	.428	7.209
3	Biacromial diameter	.501	6.908
4	Sitting height	.520	6.820
5	Arm circumference	.769	5.111
6	Abdomen circumference	.818	4.593
7	Calf circumference	.853	4.171
8	Chest circumference	.868	3.972
9	Suprailiac fatfold	.874	3.888

Table III (continued)

Males age 4-6 years; N = 623

Step	Variable added	R	Standard error of estimate
0	----	----	13.189
1	Bicristal diameter	.139	13.070
2	Stature	.336	12.439
3	Sitting height	.380	12.225
4	Biacromial diameter	.410	12.065
5	Calf circumference	.516	11.341
6	Chest circumference	.564	10.942
7	Abdomen circumference	.575	10.855
8	Arm circumference	.580	10.809
9	Suprailiac fatfold	.580	10.820

Females age 4-6 years; N = 583

Step	Variable added	R	Standard error of estimate
0	----	----	10.013
1	Bicristal diameter	.346	9.402
2	Stature	.377	9.290
3	Sitting height	.480	8.806
4	Biacromial diameter	.524	8.561
5	Abdomen circumference	.788	6.196
6	Calf circumference	.849	5.313
7	Arm circumference	.871	4.945
8	Chest circumference	.879	4.806
9	Triceps fatfold	.882	4.759

Table III (continued)

Males age 7-10 years; N = 900

Step	Variable added	R	Standard error of estimate
0	----	----	10.173
1	Biacromial diameter	.214	9.943
2	Stature	.396	9.352
3	Sitting height	.444	9.130
4	Bicristal diameter	.466	9.020
5	Arm circumference	.746	6.795
6	Abdomen circumference	.783	6.351
7	Calf circumference	.796	6.179
8	Chest circumference	.804	6.072
9	Triceps fatfold	.804	6.078

Females age 7-10 years; N = 843

Step	Variable added	R	Standard error of estimate
0	----	----	15.291
1	Bicristal diameter	.103	15.218
2	Stature	.255	14.802
3	Sitting height	.334	14.435
4	Biacromial diameter	.348	14.366
5	Arm circumference	.642	11.760
6	Abdomen circumference	.663	11.487
7	Calf circumference	.671	11.378
8	Chest circumference	.677	11.310
9	Subscapular fatfold	.679	11.288

Table III (concluded)

Males age 11-14 years; N = 884

Step	Variable added	R	Standard error of estimate
0		----	9.549
1	Biacromial diameter	.150	9.447
2	Stature	.368	8.888
3	Sitting height	.498	8.292
4	Bicristal diameter	.527	8.135
5	Arm circumference	.871	4.697
6	Abdomen circumference	.907	4.041
7	Calf circumference	.931	3.507
8	Chest circumference	.938	3.332
9	Triceps fatfold	.942	3.222

Females age 11-14 years; N = 764

Step	Variable added	R	Standard error of estimate
0			12.779
1	Biacromial diameter	.218	12.479
2	Stature	.352	11.978
3	Sitting height	.445	11.464
4	Bicristal diameter	.479	11.249
5	Arm circumference	.799	7.717
6	Calf circumference	.830	7.154
7	Abdomen circumference	.850	6.758
8	Chest circumference	.859	6.582
9	Suprailiac fatfold	.864	6.462

followed by calf circumference, were the most important.
None of the three fatfold thicknesses was a significant
contributor to the overall regression in the population
examined.

DISCUSSION

On the basis of the statistical evidence presented
in this paper, the prediction of relative body weight is
improved if the shape of skeletal frame is considered.
The subjects with a wider bony frame and a greater proportion of trunk length to leg length will have a higher
relative body weight for the same stature. In populations that vary in the shape of the skeleton, it may be
desirable, therefore, to record all the four measurements
of skeletal frame. In the examined population, however,
the regression analysis indicated that sitting height
was highly correlated with body height. Therefore, it
could be eliminated from the measurements of skeletal
size for practical purposes. The biacromial and bicristal
diameters were slightly more highly correlated with body
weight than with height but their contribution would not
significantly improve the standard error of the estimate
of body weight (Table IV).

The introduction of the seven variables related to
"muscle and fat" significantly reduced the standard error
of measurement in the estimation of relative body weight
(Table III). Of the four circumferences, the abdomen
circumference is correlated highly with chest circumference and need not be measured separately. Chest circumference, on the other hand, is well correlated with
arm circumference. Because chest circumference cannot
always be conveniently measured, due to the difficulty
of standardizing this measurement in small children and
in girls after puberty, arm circumference and calf circumference would be the measurements of choice.

In regard to the measurement of subcutaneous fat,
the analysis has shown that fatfold measurements are
highly correlated with arm circumference. In fact, these
measurements contributed only moderately to the prediction
of relative body weight. However, this does not mean that

TABLE IV

The Contribution of Height, Biacromial and Bicristal
Diameter to the Estimate of Body Weight. Stepwise
Regression Analyses and Multiple Correlations Obtained
at Each Step

Sex	Age (years)	Mean weight	Step 1 Height		Step 2 Height, biacromial and bicristal diameter	
			r	SE of estimate (kg)	R	SE of estimate (kg)
M	1- 4	8.36	.913	0.690	.935	0.603
F	1- 4	7.82	.886	0.702	.908	0.640
M	5- 9	15.73	.797	1.982	.821	1.880
F	5- 9	15.18	.632	3.143	.640	3.119
M	10-14	25.96	.862	2.790	.889	2.522
F	10-14	25.82	.781	3.650	.808	3.441
M	15-19	40.56	.885	4.264	.911	3.769
F	15-19	42.81	.826	5.377	.864	4.805

measurements of subcutaneous fat should not be used,
because the data presented in this paper were obtained
on populations in which a tendency toward obesity was
rare.

In many populations there is a tendency for later
generations to be taller, in comparison with earlier
generations at corresponding ages. This has been interpreted as an effect of improved ecological factors,
notably nutrition. Thus, the measurement of body height
in an ethnically homogenous population provides two types
of information. First, a shorter stature would, in a
cross-sectional study, indicate past malnutrition or
chronic existing malnutrition and further, if height
measurements were taken regularly over a period, the
observed increment would indicate changes in the ecology
and nutrition of the community.

A relatively shorter stature, accompanied by inadequate weight for height, would delineate further the
chronicity of the present malnutrition. A shorter stature
accompanied by normal weight for height would indicate
past malnutrition that has been corrected sufficiently
to have allowed an improvement in the development of the
soft tissues but has not been of sufficient duration to
influence the growth of the skeleton. Therefore, in
addition to height and weight measurement, other parameters that are useful for the evaluation of nutritional
status should be recorded to assess trends in the community.

Improvements in ecological factors do not influence
body height alone. Urban children show changes in body
build in addition to becoming taller. The sitting height/
standing height index and particularly the linearity-
laterality index (biacromial + bicristal diameter/height)
change also. Urban populations seem to be changing from
the more "squatty" appearance of the rural population
to the more "lanky" appearance of a control, but ethnically
different, tall population (Tables V and VI). Therefore,
it would be desirable to have more data from population
studies in regard to these indices to help explain whether
the so-called ethnic differences in body build are due to
ecological factors.

TABLE   V

The Biacromial + Bicristal Diameters/Stature in Three
Groups of Boys. Rural (R) and Urban (U) Groups Belong
to the Same Ethnic Background Whereas the Control (C)Is
the Tallest but Ethnically Different Group

Age	Rural		Urban		Control		Statistical significance		
(years)	$\bar{X}$	s.d.	$\bar{X}$	s.d.	$\bar{X}$	s.d.	R:C	R:U	U:C
1	40.6	2.1	41.3	1.8	40.4	2.2		*	*
2	39.4	2.0	40.5	1.5	38.4	1.5	***	**	***
3	38.8	1.9	38.8	1.7	37.8	1.9	**		**
4	38.8	1.5	37.6	1.3	37.8	1.5	***	***	
5	38.3	1.6	36.9	0.8	37.3	1.7	**	***	
6	38.2	1.5	36.1	1.5	37.2	1.4	***	***	***
7	37.7	1.6	36.6	0.8	36.4	1.7	***	***	
8	37.5	1.1	36.4	1.0	36.3	1.1	***	***	
9	37.7	1.3	37.1	1.8	36.8	1.4	***		
10	37.3	1.4	36.3	1.5	36.8	1.4		***	
11	37.0	1.6	36.1	1.6	36.3	1.4	**	**	
12	37.2	2.2	36.0	1.4	35.9	1.1	***	***	
13	37.1	1.5	36.3	1.0	35.7	1.4	***	***	*
14	37.0	1.4	36.4	1.3	36.1	1.2	***		
15	37.4	1.1	37.1	1.5	36.6	1.4	**		
16	38.6	1.5	37.5	1.3	37.1	1.8	**	*	
17	38.5	1.3	37.5	1.3	37.3	1.0	***	***	
18	38.8	1.7	37.4	1.1	37.8	1.5	***	***	

$\bar{X}$ = arithmetic mean;   = standard deviation;
* = $p < 0.05$; ** = $p < 0.01$; *** = $p < 0.001$.

TABLE VI

The Biacromial + Bicristal Diameters/Stature in Three
Groups of Girls. Rural (R) and Urban (U) Groups Belong
to the Same Ethnic Background Whereas the Control (C)
Is the Tallest but Ethnically Different Group

Age	Rural		Urban		Control		Statistical significance		
(years)	$\bar{X}$	s.d.	$\bar{X}$	s.d.	$\bar{X}$	s.d.	R:C	R:U	U:C
1	40.1	2.1	41.4	1.9	40.3	1.6		***	*
2	39.3	1.8	40.0	1.5	38.8	1.8		**	**
3	38.8	1.8	38.5	1.9	37.6	1.4	***		*
4	38.4	1.9	37.4	1.2	37.1	2.0	***	***	
5	38.2	1.6	36.7	1.0	37.1	1.5	***	***	
6	37.9	1.5	37.0	1.5	37.1	1.5	**	**	
7	37.9	1.5	36.5	1.3	36.4	1.2	***	***	
8	37.7	2.4	36.8	0.9	36.1	1.5	***	**	**
9	37.6	1.4	36.4	1.6	37.2	1.2		**	*
10	37.7	1.6	37.3	2.8	36.9	1.3	***		
11	37.2	1.3	36.4	1.3	36.5	1.4	**	*	
12	37.5	1.3	36.9	1.4	36.7	1.2	***	*	
13	38.0	2.3	37.4	1.1	37.2	1.3	**		
14	38.1	1.8	37.8	1.3	37.1	1.2	***		
15	38.1	1.4	37.6	1.7	37.9	2.1			
16	38.8	1.6	38.2	1.4	37.5	1.3			*
17	39.6	1.7	38.3	1.7	37.9	1.5	*		
18	39.4	1.4	38.5	1.3	38.6	1.7		*	

$\bar{X}$ = arithmetic mean; S. D. = standard deviation;
* = $p < 0.05$; ** = $p < 0.01$; *** = $p < 0.001$.

CONCLUSION

On the basis of the data presented here, the anthropometric measurements that should be included in epidemiological surveys for the assessment of nutritional status are weight, height, biacromial and bicristal diameters, arm, chest and calf circumferences and the measurement of at least one fatfold. The minimum number of measurements that could be taken conveniently by less specialized personnel includes weight, height, arm circumference and triceps fatfold. Under the circumstances in which the recording of even weight and height is not feasible but a rapid information survey is required, arm circumference as a single anthropometric measure can provide satisfactory information on nutritional status of the subject between the ages of one and 18 years. The final choice should be based on statistical analyses similar to those reported here and on associations between anthropometric data and the function of individuals.

To study trends of change in environmental factors, in a community, the recording of anthropometric indices that are descriptive of the body shape, in addition to periodic measurements of body height during growth, would be desirable.

Acknowledgment

The authors wish to express their appreciation for the assistance given by Mr. H. Dixon, Health Statistical Methodology, WHO, Geneva.

REFERENCE

1. Jelliffe, D.B.: The Assessment of the Nutritional Status of the Community. WHO Monogr., Ser. No. 53. Geneva, 1966.

CROSS-SECTIONAL VERSUS LONGITUDINAL STUDIES

Francis E. Johnston

From the Department of Anthropology, University of Pennsylvania, Philadelphia, Pennsylvania USA

The conceptual basis for this consideration of research design in the interrelationships between physical anthropology, as a discipline, and nutritional status, as a parameter, rests on the assumption that the morphology and body composition of the individual, and the biomass of a population, are functions of the quantity, the quality, and the utilization of the diets of its members. While there are other determinants that contribute to the above features, and that prevent a simplistic application of resulting data, the use of measurements as indicators of the nutritional adequacy of a group has been shown to be of sufficient utility[1-3] to suggest that such an approach be significantly informative under appropriate conditions.[4]

Several kinds of measurements are made, each with a sufficient developmental and/or physiological basis to suggest that nutritionally-relevant information is contained in the resulting variability. Thus, the anthropometric evaluation of body fat may be used as rather specific evidence of a caloric excess over that required by the body as a result of its metabolism, its activity patterns, and the like. At the same time, more global dimensions such as stature, weight, and among children and youths, the degree of biological maturation, serve as more generalized estimates of nutritional status.

Certain measurements will be common to all age
groups, others applicable to adults, and still others
to the young. The proper design of a study will ration-
ally fit the dimensions selected to the parameters to
be estimated, within the context of the age and sex of
the sample being examined. Only with such an approach
will there be any hope of the emergence of useful
results.

Another important aspect of design, too often
incompletely considered, or even ignored, consists in
determining whether a single or a series of follow-up
observations should be made of the individuals who comprise the sample. That is, whether to utilize a cross-
sectional or a longitudinal design. While it is true
that the potential information to be gained from a longitudinal approach increases exponentially over that
gained from a cross-sectional one, it is likewise true
that the dangers, the pitfalls, and the sources of
experimental and analytic error are also greater and
more subtle. The investigator who believes that to
repeat measurements on his subjects will solve all of
his or her problems is well on the way to committing a
serious error.

It is to the immediately preceding that this paper
will address itself. I will discuss the applications
of these two designs to problems studied by human biolo-
gists with special reference to the determination of
nutritional status. To be emphasized will be the sorts
of data that may be obtained by each design, and their
utility in the context of the questions being asked.
Also to be covered is the type of errors that is involved with each of the designs.

GENERAL CONSIDERATIONS OF CROSS-SECTIONAL AND LONGITUDINAL DESIGNS

The basic element of any research design that contains an experimental component is change. In the clas-
sic laboratory design, this refers to the change (or
lack of it) associated with the imposition of some

experimental treatment, other relevant variables being controlled. A second sample, identical except that no experimental treatment is imposed upon it, provides information on behavior in the absence of manipulation. Thus, if the diets of a sample of rats developing in a laboratory are experimentally altered at appropriate ages, the relationship between brain weight and body weight will differ from that of another sample identical except for dietary alteration.[5] The above design is, of course, essentially longitudinal and virtually no laboratory scientist would consider conducting a study without pre-treatment and post-treatment observations, to say nothing of suitable controls.

In the world of the natural scientist, however, such elegantly simple designs are rarely, if ever, feasible and the researcher must at the outset decide whether he or she is going to attempt to measure change which <u>will be associated</u> with some experimental variable (that is, a longitudinal design) or to measure a variable which <u>has been affected</u> by the experimental factor (that is, a cross-sectional design).

Within the discipline of physical anthropology, the choice of a particular design has all too often been treated in a cavalier fashion, the final selection being dictated by other considerations. On the other hand, among epidemiologists, the elements that determine the choice of methodological approach have been the subject of many papers and discussions. By their very nature, longitudinal studies are called <u>prospective</u>[6] because they permit the measurement of some change that it is hypothesized will occur. At the same time, cross-sectional studies are called <u>retrospective</u> or <u>prevalence</u> studies because they may only be used to measure the current status of the population, with any controlling mechanisms being elicited by the reconstruction of what has happened.

With respect to the study of child development, Tanner[7] has succinctly characterized the data resulting from these two approaches as being either "distance" or "velocity" in nature. Since growth is "a form of

motion," cross-sectional studies provide information on
the distance "travelled" by an individual, or sample,
up to the time when observed. Since longitudinal data
permit the calculation of change, they provide information, with respect to time, on the rate of change per
unit of time, or the velocity.

Thus, the basic conceptual frameworks of longitudinal and cross-sectional designs differ in that they
ask essentially different questions. The latter seek
to describe a parameter in terms of what has happened
and to make inferences from a reconstruction of the
natural history of some phenomenon that is presumably
known. To the laboratory scientist such an approach is
untenable; to the natural scientist it may be the only
one possible and, in view of the questions asked, the
problem studied, and the resources available, may be
the more desirable of the two.

The longitudinal design seeks, not to describe a
parameter in the static sense, but to ascertain changes
in it which are then associated with relevant variables.[8] The study of a single cohort, or of a group of
cohorts, across a range of time, and the relationship
of change in some parameter measured in the cohort,
form the basis of the "epidemiological experiment"[9]
and, for the natural scientist constitute the analogue
of the laboratory experiment. Longitudinal designs,
regardless of the ages of the subjects comprising the
sample, are not without serious theoretical and practical difficulties and these difficulties are such that
they have, more often than one wishes to remember, cre-
ated more problems in their answers than existed in the
questions being asked.

Were there no change in the natural world there
would be no need to consider these two basic research
designs. However, change is integral to life and our
problem becomes not only one of discussing the nature
of change, but of developing ways of analyzing it with
respect to pertinent variables; as Campbell[10] has
phrased it, of moving from description to experimenta-
tion in such a way as to interpret "trends as quasi-
experiments."

CROSS-SECTIONAL STUDIES

The traditional design of much of the research that has been conducted in physical anthropology and human biology is the cross-sectional one. A given sample is selected and a single set of observations made. This results in one array of data. From such an array, what kinds of questions may be answered?

The first, of course, rests in the description of the population. Without reference to experimental treatment or to causative factors, it may be useful simply to describe the morphology of the population that has been sampled. The scientific literature abounds with descriptive studies that make no attempt to relate their findings to other variables, or perhaps only to such factors as age or sex. Interpretations of such data reveal the basic weakness of cross-sectional studies; that is, the frequent failure to control contravening variables. Even if the investigator wishes merely to quantify differences between males and females, he or she must know something about other conditions that may affect these differences. For example, nutritional deficiencies may alter male and female morphology differentially and instead of quantifying differences determined by gender, the investigator is actually recording sex differences as reflected in the response to suboptimal nutritional intakes.

Cross-sectional studies are used frequently to infer the operation of causative factors in ways that mimic the experimental approach. A study sample, observed but once, is compared to another sample or to a normative standard, presumably similar in all significant aspects except for the variable studied. Differences in the measurements are attributed to the operation of the variable. The assessment of the nutritional status of a population, utilizing the anthropometry of a sample studied cross-sectionally, provides an example of this approach. The biometric profile of the study sample is assumed to have been derived from the cumulative nutritional history experienced by the individuals. Hereditary variation, idiosyncratic differences, or

other environmental determinants that also differ between the study and the control population, are assumed
either to be inoperative or to contribute no significant
bias.

Thus cross-sectional studies are inherently weak
in that they are retrospective; that is, whatever conditions that affected the measured variables have done
so in the past and must be reconstructed to the best of
the investigator's ability. What can one conclude about
the nutritional status of a population living at high
altitude, with consequent differences in oxygen availability and ambient temperature, compared to a control
sample living near sea level in a subtropical environment and demonstrably different in their genetic structures? Clearly the investigator ought to know not only
the study sample in a detailed manner, but must ensure
that the study and control groups are as nearly similar
with respect to significant factors as possible.

The estimation of the effects of determinants from
a retrospective study is, of course, not always possible and inconsistencies, haziness, and even controversy
have frequently been associated with the findings from
them. The problem of migrant populations in radically
different environments provides an example of the problems associated with a cross-sectional design. Significant differences between the size and shape of
migrants and sedentes have been shown in a number of
studies spanning half a century.[11-14] Usually these
differences are attributed to changes brought about by
environmental variations and frequently to improved
health and nutritional conditions affecting the migrants
in their new homeland. However, other scientists have
questioned the comparability of migrants to sedentes,
because the psychological and biological constitutions
of the two groups may differ. Migrants may differ when
they leave, or they may react differently to a new
environment. The point is that retrospective studies
can never answer this question completely.

Another limitation of a cross-sectional approach
to the estimation of change is found when the change

studied is developmental, or when a change is inferred
from a difference between two comparable groups. In
some cases the data are acceptable. Thus, when Dreizen
and his associates[15] compare well-nourished and undernourished Alabama girls, as shown in Table I, the two
groups seem similar enough to allow acceptance of the
view that the delays in the undernourished girls resulted from deficient diets.

On the other hand, the measurement of change
through time in growing children, frequently studied
with a cross-sectional design, is subject to considerable error. The source of error rests in the fact that
different children are measured at different ages and
the mean growth from one age to the next is estimated
as the difference between the means of the measurements
for the two age groups. The children comprising each
age group, being but a sample of that cohort, are subject to a sampling error that is independent of that
associated with adjacent age groups. Since the sampling error of a difference between sample means is

TABLE I

Comparison of developmental parameters in 30
well nourished and 30 undernourished Alabama
white girls[15]

Group	Height at 12.4 years (cm)	Height at 14.4 years (cm)	Age at Menarche (years)
Well nourished	154.9	161.5	12.4
Undernourished	145.9	158.0	14.5
Difference*	9.9	3.5	2.1

*All differences are statistically significant.

twice that of the error of either mean, what may be
interpreted as growth from a cross-sectional curve may
in fact reflect uncontrollable error inherent in the
design of the study.

For example, the distance curve of attained height
against age, shown in Figure 1, and taken from the
unpublished data of W. M. Krogman, is constructed from
cross-sectional data on Philadelphia Negro children
and youths from 7 through 17 years of age. The curves
are, as expected, generally smooth, though some fluctuation may be observed. Although the sampling error
of any age group mean may be estimated within acceptable levels of probability, the specific effects remain
unknown in particular groups. For example, the difference between the means of the females comprising the
14- and the 15-year age groups is less than that between

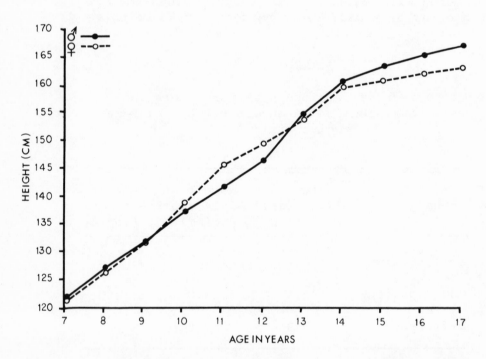

Figure 1. Means for height in Philadelphia Negro children. Unpublished cross-sectional data of W.M. Krogman.

the 15- and 16-year age groups. Are we to assume that
there is, among individual Negro females, deceleration
of growth in height in the first period followed by a
subsequent acceleration, or do we attribute this to
sampling error? Clearly, any interpretation requires
not only a general knowledge of human growth, but also,
the application of some subjectivity in accepting or
rejecting results as valid.

Larger sample sizes and attention to the details
of sampling can reduce the associated error and ensure
that the calculated group means approximate more closely
those of the population. However, even under the most
ideal conditions, one may not always accept the results
of cross-sectional studies completely. For example,
the Health Examination Survey of the United States Public Health Service has, since 1962, conducted a comprehensive survey of the health and growth of American
children. This survey, which covered the population
aged 6 through 17 years by 1970, was based on the most
sophisticated probability sampling technique utilized
by the United States government and involved measurements of approximately 16,000 children. If one examines the cross-sectional curves for weight,[16] it will
be noted that the mean weight of 17-year old females is
0.49 kg (1.1 lb) less than that of 16-year olds. Is
this a significant reduction suggesting, e.g., a tendency for weight control among adolescent females, or
are we to dismiss it as sampling error, even in the
face of such statistically and demographically sophisticated techniques? If the sampling error can be controlled to a reasonable degree, then the differences
between means will give us the average amount of
growth. All else being equal, the difference between
means of two samples equals the mean difference of
individuals of a single cohort measured serially.

Cross-sectional data are again severely limited if
used to study the growth of youths during their adolescent years. Because individuals enter their growth
spurt at any age within a range spanning four or more
years, and because the shape, intensity, and duration
of this spurt are related to the age of onset, the

grouping of individuals by chronological age categories
causes an "out-of-phase" effect that will systematically
distort estimated growth increments. Beginning with
Boas,[17] and elaborated by Shuttleworth[18] and Tanner,[7]
investigators have consistently pointed out this fact
and demonstrated that a cross-sectional, chronological
age-based treatment will grossly underestimate the magnitude of the spurt in individuals. This can be seen
in Table II, where the differences between successive

TABLE II

Comparison of differences
between means for height, age
grouped, and mean annual
velocity, PHV-grouped in
Guatemalan males[19]

Cross-sectional treatment Age-grouped		Longitudinal treatment PHV-grouped*	
Mid-point	Difference (cm)	PHV group	Difference (cm)
11.0	4.7	-3	4.9
12.0	5.8	-2	4.9
13.0	7.2	-1	6.0
14.0	4.4	PHV	9.5
15.0	7.9	+1	5.7
16.0	2.7	+2	3.3

*PHV-grouped means grouped
in relation to the age interval before (-) or after (+)
peak height velocity. In
this sample, the average age
of peak height velocity was
13.62 years.

mean values of chronological age groups, for the adolescent years, are presented for a sample of well-nourished
Guatemalan children.[19] These may be compared to the
actual mean increments calculated for individuals and
expressed, not as chronological ages, but as years before
and after their peak adolescent velocities in height.
The systematic underestimation is clear.

These same problems apply to the cross-sectional
monitoring of nutritional status across time. For example, an investigator may take an appropriate series of
anthropometric measurements on a sample of individuals
of the same age or ages and use these dimensions to
estimate the nutritional status of the population. Suppose these same measurements are repeated five years
later on another sample of the same age distribution as
the first. To what extent may significant differences
between the two groups in these measurements indicate
changes in nutritional status? Systematic bias in sampling may affect the dimensions without regard to nutritionally relevant parameters. The investigator must be
assured that migrations, changes in social mobility, or
other factors have not affected the selection of individuals such that another population, different from the
original, has in fact been sampled. Even when systematic
bias has been controlled, the specter of random sampling
error must be considered in a cross-sectional design.
Differences between samples may have nothing to do with
changes in nutritional status but may have occurred
simply through sampling error. This may occur even with
the best sampling design. The above criticism imposes
severe limitations upon the utility of a cross-sectional
design and upon any resultant data, but particularly
where change is a relevant variable to be measured.

On the other hand, cross-sectional designs are not
without some advantages, particularly in cases where the
limitations are realized and where their contributions
to experimental error are controlled as far as is
possible. First and foremost, they are relatively quick
to conduct and, as such, require far less expenditure of
time, energy, and funds. Wherever the potential gain
relative to expenditure must be considered, cross-

sectional studies must not be summarily rejected. The
unfortunate history of some of the best known longitudinal studies bear testimony to the ease with which
data collection can become an end in itself, and to the
very long delay between the initiation of such a study
and the emergence of findings that exploit the unique-
ness of longitudinal data. A cross-sectional study may
be begun, completed, analyzed, and the results published
in the time it takes to measure change over a single
time interval in a prospective study.

Comprehensive cross-sectional studies will reduce
the impact of sampling error. Scott's study of London
school children[20] and the United States Health Examination Survey[21] are noteworthy examples of the reduction
of sampling error to a tolerable minimum in a cross-
sectional study though, as noted above, even such
studies are not without interpretive problems.

Careful analysis of differences between age groups
may produce significant results, where sample sizes are
large and carefully drawn. Though its magnitude could
not be determined, the timing of the adolescent growth
spurt among American youths was estimated, through
painstaking analysis, from the data of the Health Examination Survey.[16]

LONGITUDINAL STUDIES

The objections to a cross-sectional design are met
by a longitudinal protocol. By its very nature, such a
design is prospective; i.e., the sample is selected in
advance and the change is observed through measurements
across a time range. Associations with relevant variables may be determined directly. For example, if one
is studying the role of seasonal variation in nutrient
availability, a sample of individuals may be examined
over the course of a season, or seasons, and any changes
noted.

One immediate advantage of such a design is that
the sample may serve as its own control. While it may

sometimes be useful or even necessary to employ other
samples as control, i.e., not subject to the "experimental treatment," the dependence upon some other reference standard may be alleviated and a source of error
eliminated.

In studying age changes, growth may be estimated
as the difference between a dimension measured on two
successive occasions and the investigator may then com-
pute the velocity from the formula

$$\text{velocity} = \frac{x_2 - x_1}{t_2 - t_1}$$

where x indicates the value of the same dimension measured on two occasions, t_2 and t_1. These velocities
may be treated collectively and group curves constructed. Because of their sensitivity, such curves
have been shown to be more useful in clinical situations
than the more traditional distance curve. The response
of an individual to therapy, or to an altered nutritional state, is evidenced more dramatically in his or
her rate of growth than in attained size.

Thus the predominant advantage of the longitudinal
design, and one that cannot be matched by a cross-
sectional approach, is that the former permits the
quantification of individual change. Any change estimated by the differences between the means of cross-
sectionally studied groups is the mean change. Only
with longitudinal studies can individual variability
about this mean be determined. Thus, in our longitudinal study of Guatemalan children,[19] a cross-sectional
treatment of the data and a longitudinal treatment will
both show that the average amount of growth in boys between 7.5 and 8.5 years is 5.5 cm; however, only the
longitudinal treatment will show that the standard deviation of increments about this mean is 1.2 cm.

The advantages of a longitudinal study are therefore obvious, particularly when viewed relative to the
disadvantages of the cross-sectional method. However,

one should not jump to the conclusion that a prospective study provides all of the answers to existing
problems, nor should one remain unaware of the problems
inherent in such a design.

Longitudinal studies are expensive and time-
consuming. A single examination of a subject may be a
relatively simple procedure; two examinations of that
subject over a single time interval requires control
over subject availability. The effort necessary to
obtain n examinations over n - 1 intervals is alarmingly
greater. Investigators should not undertake a longitudinal study without full awareness of the administrative and logistic details necessary to avoid an
intolerable attrition rate in the sample.

Longitudinal studies cannot be as comprehensive
across the range of the population as can cross-sectional
ones. They are intensive rather than extensive and
must focus upon a more restricted sample. Since the
sample size, in terms of individuals, will be smaller
and since, in all probability, it will be drawn from a
segment of the population more easily followed across
time, the probability of systematic sampling bias will
be increased. Cross-sectional studies are subject to
sampling error within each age or treatment cohort, but
this error is variable in its effects and randomized
throughout the total sample. On the other hand, because a longitudinal sample consists of a given cohort
followed through time, sampling error associated with
subject selection will be repeated in every group and,
rather than being random, will systematically bias the
results.

The effect of this systematic bias is most acute
in the area of child development, where a given group
may be followed for 10 to 20 years. Despite the recognition of its existence, the quantification of its
effect has generally been lacking. One noteworthy
exception is to be found in a comprehensive study of
the growth of Finnish children from birth to 10 years.[22]
In this study, Tiisala and Kantero[23] isolated possible
bias in the longitudinally-followed component. They

found that a "disproportionately greater number" of large boys had apparently been included in the sample and any clinical use of the data must take this into account.

In a longitudinal design the sample is selected prospectively from among a population susceptible to some particular agent. If only a portion of the total population is actually affected by the agent, it is possible that only a small percentage of the sample selected for serial monotoring will be among the affected. The researcher will be in the unfortunate position of having studied the wrong individuals.

Longitudinal studies, just as cross-sectional ones, are also beset by methodological and analytical difficulties. It has become widely known that the time interval between observations is of crucial importance; Linden[24] has demonstrated that this interval effect may distort the data in ways uncontrollable by the investigator. The estimated intensity of a growth spurt may vary by as much as 50 percent depending upon the time interval between measurements; this is confounded by the investigator's not knowing if the first examination actually, as is highly desirable, coincided with the onset of the spurt. The same consideration holds true for epidemiological variables and "great care must be taken in defining the duration variable of interest."[25]

Even with the increments quantified and expressed as rates or velocities, the analytic procedures are not always clear. How does one analyze increments? Since many have zero as a lower limit, one would suspect that they would be skewed to the right and therefore suggest either transformation or non-parametric techniques. The exchange between Meredith[26] and Garn and Rohmann[27] over this matter did not completely clarify the issue, though Garn and Rohmann's conclusion that "fully symmetrical incremental distributions are the rarity" seems justified.

ATTEMPTS AT RECONCILIATION OF THE PROBLEM

There have been some attempts to reconcile the problems involved in selecting one of these designs. The one most often seen is based upon the fact that a pure longitudinal study is almost impossible to carry out due to attrition, cost, and the like. One solution is the mixed longitudinal approach, in which there is a cross-sectional and a longitudinal element. The latter will be examined serially, and may consist of individuals from a range of age and sex cohorts; the former may be measured at the outset or added to periodically. The two designs will complement each other; one may be used to test the presence of error in the other. In a very thoughtful paper, Tanner[28] has suggested that the most accurate representation of a given size parameter is to be obtained by utilizing both longitudinal and cross-sectional elements, though the computation of the estimate is probably too involved to appeal to most research workers.

MEASUREMENT ERROR

One of the most important, but one of the least well analyzed aspects of any anthropometric study, involves the error of measurement. While any analytic technique is subject to an observer error, the measurement of a linear dimension, a soft tissue thickness, or a surface contour on the human body are all particularly sensitive. There are many reasons for this, among which are:

1. Inadequate technician training. Many investigators take an unjustifiably casual attitude toward anthropometry and consequently fail to train technicians adequately, to establish procedures for ongoing monitoring, or to quantify the contribution of error to the total variation. Without such procedures, the findings of any study cannot be accepted with confidence. In preparation for this section, four textbooks on anthropology were consulted; none of them presented any discussion of error of measurement, or its quantification, despite elaborate descriptions of the techniques themselves.

2.   Inherently large errors due to units of measurement. Some dimensions display small values relative to the units in which they are expressed. Thus a
fatfold may be, e.g., 15 mm thick. If its thickness is
measured with a Lange caliper, calibrated in millimeters,
it will be possible to measure it to the nearest 0.5 mm.
A measuring error of one "unit" (i.e., 0.5 mm) will
result in a three percent error.

3.   Inherently difficult measurements. Some dimensions are more difficult to measure than others and,
of necessity, have larger error components. Among these
are fatfolds and sitting height.

The errors associated with anthropometry are of
two types: variable and systematic. Variable errors
are those in which a persistent tendency is not noted.
They do not affect the mean of a distribution but will
increase measures of variability about it. Systematic
errors are characterized by a persistent direction
relative to the true value; thus in intraobserver error,
the second measurement might always be lower than the
first; or in interobserver error, technician B might
systematically obtain a higher reading than technician A. Systematic errors do not affect the variance
of a distribution as much as the mean.

The type of error more easily tolerated will play
a part in dictating the choice of a longitudinal or a
cross-sectional design. In a longitudinal study, em-
phasizing change, systematic error is not as crucial,
since the difference between measurements is the parameter of interest. Suppose that an individual measurement, x, is repeated at visits 1 and 2 by a single
observer, a, who will systematically obtain an erroneously large measurement because of some bias, e. Suppose that this measurement is made on the same individual at the same times by observer b, with a systematic
bias in the opposite direction, i.e., the measurement
is erroneously small.

Thus,

$$x_1 + e_a \neq x_1 - e_b, \text{ and}$$

$$x_2 + e_a \neq x_2 - e_b$$

on the other hand,

$$(x_2 + e_a) - (x_1 + e_a) = (x_2 - e_b) - (x_1 - e_b)$$

That is, the calculated increments will be the same with a systematic error that affects the measurements equally. Thus a longitudinal design should employ as few technicians as possible taking measurements, thereby minimizing the effects of variable error, as well as different systematic errors by different investigators. If, in the above example, observer b had measured at visit 1, and a at visit 2, then,

$$(x_2 + e_a) - (x_1 - e_b) = x_2 - x_1 + (e_a + e_b)$$

i.e., the increment would have been in error by the sum of the two observer errors.

The use of more than one observer cancels out systematic bias in cross-sectional studies, but increases the variance due to the introduction of variable error. In the U.S. Health Examination Survey, replicate examinations were made for the purpose of quantifying and reporting error. The median intraobserver error in triceps fatfold measurement (77 replicates) was 0.5 mm.[29] For interobserver error among 11 technicians (224 replicates) the median value was 1.5 cm. Thus while any systematic bias attributable to a single observer was eliminated by using several technicians, a procedure made necessary also by administrative and logistic considerations, the variable error was increased three-fold (0.5 to 1.5). No firm recommendations can be made concerning technician number, technician error, and study design. All factors should be considered, including the purpose of the investigation, in making the final choice. Appendix III of Johnston et al.[29] presents many of the factors relative to the

error of measurements in longitudinal and cross-
sectional designs.

APPLICATION TO NUTRITIONAL EVALUATION

Jelliffe[2] has discussed epidemiological aspects
of longitudinal and cross-sectional designs in nutritional assessments. His major points depict adequately
the advantages and disadvantages of both and he has
concluded that more longitudinal studies are needed.
Seasonality has been the downfall of many human biological studies; nutritional assessments have not been
spared. Repeated studies across ecological and cultural cycles will reduce errors inherent in one-time
observations.

If age changes are significant, both cross-
sectional and longitudinal studies are necessary to
establish trends. Cross-sectional studies must be
truly extensive and must employ appropriate sampling
techniques to allow generalization. Longitudinal studies must be equally intensive and must involve the most
sophisticated techniques of design and analysis, involving curve fitting and multivariate models[30,31] to
ensure that the intrinsic value of the data they provide will be exploited fully.

REFERENCES

1. Brožek, J., (Ed): Body Measurements and Human
    Nutrition. Detroit, Wayne Press, 1956.

2. Jelliffe, D.B.: The assessment of the nutritional
    status of the community (with special reference
    to field surveys in developing regions of the
    world). WHO Monog. Ser., 53:1, 1966.

3. Kelsay, J.L.: A compendium of nutritional status
    studies and dietary evaluation studies conducted
    in the United States, 1957-1967. J. Nutr., 99,
    Suppl. 1:119, 1969.

4. Keet, M.P., Hansen, J.D.L. and Truswell, A.S.:   Are
   skinfold measurements of value in the assessment
   of suboptimal nutrition in young children?   Pediatrics, 45:965, 1970.

5. Dobbing, J.:   Undernutrition and the developing
   brain, p. 241.   In: Himwich, W.A. (Ed) Developmental Neurobiology.   Springfield, Ill., Charles
   C Thomas, 1970.

6. Holland, W.W.:   Principles of study design, p. 45.
   In: Holland, W.W. (Ed) Data Handling in Epidemiology.   New York, Oxford Univ. Press, 1970.

7. Tanner, J.M.:   Growth at Adolescence.   Second edition.   Oxford, Blackwell Sci. Publ., 1962.

8. Goldstein, H.:   Longitudinal studies and the measurement of change.   The Statistician, 18:93,
   1968.

9. Susser, M.:   Casual Thinking in the Health Sciences:
   Concepts and Strategies of Epidemiology.   New
   York, Oxford Univ. Press, 1973.

10. Campbell, D.T.:   From description to experimenta-
    tion: interpreting trends as quasi-experiments,
    p. 212.   In: Harris, C.W. (Ed) Problems in Mea-
    suring Change.   Madison, Univ. Wisconsin Press,
    1963.

11. Boas, F.:   Changes in bodily form of decendants of
    immigrants (1910-1913), p. 60.   In: Boas, F.
    (Ed) Race, Language, and Culture.   New York,
    Macmillan Co., 1940.

12. Kaplan, B.A.:   Environment and human plasticity.
    Amer. Anthrop., 56:780, 1954.

13. Hulse, F.S.:   Exogamie et hétérosis.   Arch. Suisse
    d'Anthrop., 22:103, 1957.

14. Froehlich, J.W.: Migration and the plasticity of physique in the Japanese-Americans of Hawaii. Amer. J. Phys. Anthrop., 32:429, 1970.

15. Dreizen, S., Spirakis, C.N. and Stone, R.E.: A comparison of skeletal growth and maturation in undernourished and wellnourished girls before and after menarche. J. Pediatr., 70:256, 1967.

16. Hamill, P.V.V., Johnston, F.E. and Lemeshow, S.: Height and Weight of Youths 12-17 Years, United States. DHEW Pub. 73-1606, Ser. 11, No. 124. Washington, D.C., U.S. Govt. Printing Ofc., 1973.

17. Boas, F.: Remarks on the Anthropological Study of Children. Trans.15th Internat. Cong. Hygiene and Demog., Washington, D. C., 1913.

18. Shuttleworth, F.K.: The Physical and Mental Growth of Girls and Boys Age 6 to 19 in Relation to Age at Maximum Growth. Monog. Soc. Res. Child Develop.,4, 1939.

19. Johnston, F.E., Borden, M. and MacVean, R.B.: Height, weight and their growth velocities in Guatemalan private school children of high socio-economic class. Human Biol., 45:627, 1973.

20. Scott, J.A.: Report on the heights and weights (and other measurements) of school pupils in the county of London in 1959. London, London County Council, 1959.

21. National Center for Health Statistics: Plan, Operation, and Response Results of a Program of Children's Examinations. PHS Pub. No. 1000, Ser. 1, No. 5. Washington, D. C., U.S. Govt. Printing Ofc., 1967.

22. Hallman, N., Bäckström, L., Kantero, R.L. and Tiisala, R.: Studies on Growth of Finnish Children from Birth to Ten Years. Acta Paediat. Scand., Suppl. 220, 1971.

23. Tiisala, R. and Kantero, R.L.: Comparison of
     height and weight distance curves based on longi-
     tudinal and cross-sectional series from birth to
     ten years. Acta Paediat. Scand., Suppl. 220, 1971:

24. Linden, F.P.G.M. van der: The interpretation of
     incremental data and velocity growth curves.
     Growth, 34:221, 1970.

25. Menken, J.A. and Sheps, M.C.: On relationships be-
     tween longitudinal characteristics and cross-
     sectional data. Amer. J. Public Health, 60:
     1506, 1970.

26. Meredith, H.V.: On the distribution of anatomic
     increment data in early childhood. Amer. J. Phys.
     Anthrop., 20:519, 1962.

27. Garn, S.M. and Rohmann, C.G.: On the prevalence of
     skewness in incremental data. Amer. J. Phys.
     Anthrop., 21:235, 1963.

28. Tanner, J.M.: Some notes on the reporting of
     growth data. Human Biol., 23:93, 1951.

29. Johnston, F.E., Hamill, P.V.V. and Lemeshow, S.:
     Skinfold thickness of children 6-11 years, United
     States. DHEW Pub. 73-1602, Ser. 11, No. 120.
     Washington, D.C., U.S. Govt. Printing Ofc., 1972.

30. Deming, J. and Washburn, A.H.: Application of the
     Jenss curve to the observed pattern of growth
     during the first eight years of life in forty
     boys and forty girls. Human Biol., 35:484, 1963.

31. Murray, J.R., Wiley, D.E. and Wolfe, R.G.: New
     statistical techniques for evaluating longitudi-
     nal models. Hum. Dev., 14:142, 1971.

RELATIVE VALUES OF LONGITUDINAL, CROSS-SECTIONAL AND

MIXED DATA COLLECTION IN COMMUNITY STUDIES

Derrick B. Jelliffe

From The School of Public Health, University of
California at Los Angeles, USA

"They sought it with thimbles, they sought it with care,
   They sought it with forks and with hope,
They threatened its life with a railway share,
   They charmed it with smiles and with soap."

>   The Hunting of the Snark
>   (Lewis Carroll)

A multidisciplinary approach to anthropometric
methods of nutritional assessment is obviously indicated,
and this is endorsed by the backgrounds of those attending the present conference.

Studies or surveys to collect such data can have a
variety of purposes including:

-- the scientific and mathematical investigation of
the biological meaning of anthropometric measurements;

-- the gathering of data for the compilation of
standards or norms;

-- the evaluation of the effects of nutrition on
function;

-- the assessment of the nutritional status of the
   community and

-- the gathering of data in relation to the hunt
   for the ultimate and elusive "Snark" of making
   Scientific Nutritional Anthropometry Routinely
   Known. This is an ultimate, extremely difficult,
   but very important, goal.

The present account concentrates on assessment of
the nutritional status in the community. These comments
are practical and functional, and concerned mainly with
public health priorities in developing countries. They
are biased by personal experience, mostly in some 29
cross-sectional surveys of young children, principally
in East Africa[1] and in the Caribbean, ranging in complexity from very simple, rapid surveys, such as the
investigation of the prevalence of protein-calorie
malnutrition (PCM) of early childhood in Haiti in 1958,[2]
to full-scale investigations, such, for example, as the
National Survey of Barbados in 1969.[3]

The paper is in two parts--first, a consideration
of the comparative roles and values of longitudinal and
cross-sectional data collection in the nutritional monitoring of communities, and, second, the question of the
contributions that have been made (and could be made)
by specific research projects in assisting the development of field methods and their interpretation.

Nutritional monitoring is concerned with nutritional
assessment with a time dimension--that is, with assessment on a continuing basis in a community. This is a
major public health concern. Most nutritional work needs
to be related ultimately to this and to improvement of
the nutritional status in the community. This is a major
over-all objective.

PURPOSES

There are various purposes of nutritional assessment, and five are listed here.

1. Obviously, a major objective of nutritional
    assessment is to measure the nutritional status
    of a community, and to gauge the effects of spon-
    taneous and planned changes. This is needed
    from the public health viewpoint for the rational
    use of resources. This measurement can be for
    a community, for a specified age range, a par-
    ticular physiological group (such as pregnancy),
    etc. Sometimes, it is relatively easy, because
    in some parts of the world it is not a nutri-
    tional assessment, but a malnutritional assess-
    ment. It is a truism that the worse the nutri-
    tional status of a community, the easier the
    assessment.

2. The second purpose is to identify priority
    causes amenable to practical improvement. In
    other words, a nutritional assessment has to be
    an ecological study--in detective parlance, a
    reconstruction of the crime. Many different
    interacting factors contribute to the metabolic
    breakdown known as malnutrition. Kwashiorkor
    in Madras has a different detailed etiology
    and background from the same syndrome in Haiti.
    The simplistic concept of universal solutions
    is a figment of the imagination.

3. The third purpose of nutritional assessment
    moves from science into the wide world. This
    is to motivate and convince policy makers. For
    this, it is necessary to gain their involvement
    from quite early in the survey. Even more
    importantly, results have to become available
    to them with minimum delay. These results have
    to be understandable by the busy man who has
    many other problems. At the same time, it is
    vital to avoid the opposite extreme of producing
    over-sensational reports that too frequently
    rebound. The ultimate objective is to convince
    politicians and administrators and to alter food
    and nutrition policy.

4. A fourth purpose of a nutritional assessment is
    educational. This is perhaps away from the main

theme of this conference, but it must be mentioned. Surveys always need to incorporate an
educational component for the people working
in them--for the staff responsible, for the
administrators, and, if possible, for the population among whom the study is being undertaken.

5.   Fifthly--and it has been put last, rather pro-
     vocatively--is the collection of scientific
     information, including the testing and valida-
     tion of field methods.

Realities of Less-Developed Countries

It is necessary to reiterate the realities of life
in the majority of the world, that is in less-developed
countries, to set the context for the following comments.
There will be very limited funds available (50¢ to $2.00
per head per year for the health budget), few trained
staff, a heavy burden of work, the necessity to use
auxiliaries, communication problems (roads, distances,
etc.) and the very great likelihood of cross-cultural
misunderstandings. This is well known, but needs bearing
in mind--as it is in these circumstances that there is
a greater need for the application of simple methods of
nutritional assessment.

LONGITUDINAL DATA COLLECTION

There are obvious advantages to longitudinal collection of data and several are listed:

Malnutrition is rarely the result of diet alone,
and, even when due exclusively to abnormal intake, the
causation of the inadequate dietary is complex. There
are obviously multiple, interacting factors operating
at one time, and the only way one can really see the
full panorama laid dynamically before one--diet, infections, customs, economics, etc.--is by longitudinal
follow-up.

Another advantage of a longitudinal study is that

goodwill can be engendered by a sympathetic approach to such a study and can lead to intimate contact and knowledge of the people involved, which is quite impossible in any other way. For example, in East Africa at one time, the statistics concerning birth rates were totally wrong because those carrying out the study had little appreciation of the fact that in this community a placenta was considered as a child. The question "How many babies have you had?" obviously gave very different results from the truth.

A second advantage of longitudinal collection of data is that it gives information on the incidence of malnutrition and on conditioning infections during the whole year. Also, it covers the seasonal variations that are so important and so frequently underestimated--for example, due to changes in climate, food availability, community activities, etc.

Thirdly, information on short-term episodes is obtained. For instance, a cross-sectional survey will give little idea of the impact of measles on a community. Also, some forms of malnutrition are "short-term." Examples are infantile beri-beri and the appearance of edema.

A fourth advantage of longitudinal data collection is that the ages of all young children will, almost certainly, be known with accuracy. This is important in interpreting most anthropometric measurements. Also, if the ages are known, this provides a valuable opportunity to test the measurements that might be used in other communities where precise ages are not verifiable.

A fifth advantage of longitudinal survey of this type is that there is a considerable "spin-off" in information. Almost inevitably one finds large amounts of unexpected information becoming available due to this type of endeavor.

Three major obvious <u>disadvantages</u> will be considered.

Time is highly relevant, and covers all the activities

needed to produce results that will have impact, and that
will still be valid in the rapidly changing world.

The cost of staff, of logistics, of analysis--all
may be very large--and increasingly beyond most available
resources.

A wide variety of changes can totally disrupt, or
make very difficult, any such study--changes in the community itself and/or among the people involved in the
study. As Cicely Williams has said "any survey inevitably
alters the observer and the observed."

It is necessary to keep the staff, techniques, and
the local scene as constant as possible. Staff are particularly difficult to recruit, and to train for work as
a team over years.[4] This is where a dynamic, but non-
self-seeking leader is important. Problems will arise.
It is particularly difficult for a physician to continue
in a mostly observer role, for this is what he has been
taught not to do. This is opposed to the cultural anthropologist who has been trained to observe and not interfere with the dynamics of the local community.

CROSS-SECTIONAL DATA COLLECTION

The advantages and disadvantages are essentially
the opposite of those of longitudinal data collection.

Advantages. The cost is relatively low, although
this may be very much more than appreciated, if transport
and charges for computer analysis are estimated realisti-
cally.

These surveys are quick, they can be undertaken
in days or weeks, but again, analysis may be a major
delaying factor. Also, it is possible to implement these
studies with people who have other duties--for example,
physicians or medical students--all of whom have other
functions, but who may be able to join a relatively short-
term cross-sectional data collection program, as a teaching exercise.

An important advantage of a short-term survey is that it can have sustained motivational value for the population and for the surveyor. This is easier over a short time than over the long, often grueling years required for a longitudinal study.

Disadvantages. Many of these are obvious. Cross-sectional studies give prevalence figures only. In relation to conditioning infections alone, which are so important in a young child's nutrition, only those will be recorded which are either chronic or epidemic at the time. Likewise, in a cross-sectional survey, only those forms of malnutrition will be detected that are relatively chronic. Thus, more marasmus than kwashiorkor will be noted, even if they both happen to be recurring throughout the year in the particular community. Cross-sectional surveys give information concerning the etiology of only those conditions found at the particular time of the survey.

Anthropometric findings from a cross-sectional study are difficult to interpret because it is impossible to know whether a child is "coming or going." For example, arm circumference merely indicates that at that particular time there are a certain number of children who have so much muscle and so much fat at the midarm level. It is impossible to know whether they are on the road to recovery or are deteriorating.

The difficulty with ages cannot be over-stressed. It is a major bugbear of real life anthropometry in developing countries. Ages are not known exactly in most of the world, because in most cultures precise ages have no social significance. Plainly, this poses problems with the interpretation of anthropometric measurements in young children.

MIXED DATA COLLECTION

From a practical point-of-view, mixed data collection represents a compromise, and is usually chosen for reasons of cost and time. Frequently, the study will be based

mostly on data collected cross-sectionally, but, in
addition, available supportive longitudinal information
will be included.

It may be noted that the cross-sectional collection
of data is a very varied technique. A few years back,
what may be termed "World War I infantry attack" cross-
sectional surveys were customary--wave after wave of
investigators being involved. Nowadays sufficient funds
are not available nor, indeed, is this massive approach
considered cost-effective. What is needed is the selection of the minimum number of measurements and minimum
amount of other data that will speedily provide optimal
useful information.

Mixed collection is more usual nowadays--a cross-
sectional survey, plus longitudinal information from
various sources. For example, longitudinal information
can be obtained from health centers, which will, of
course, be biased towards the adjacent area. Also,
valuable data may be available from other health services.
For example, although hospital admission figures for
kwashiorkor and marasmus are influenced by cultural
factors, hospital policy, effectiveness and forms of
treatment, etc., they give valuable, but biased, information and should be collected. In addition, children
with severe malnutrition give clues as to the causation
of the cases that reach the health services, including,
for example, locally important conditioning infections.

Of course, all this information is statistically
suspect and far from the best that might be possible,
but cross-sectional data collected by visiting the community (by examining children, by data on food consumption and food production in the area), together with
longitudinal information, can lead to an approximate,
but rational, diagnosis of the nutritional status of the
community on which to base the selection of appropriate
preventive measures.

CONTRIBUTION OF SPECIAL RESEARCH STUDIES

Validation of anthropometric indicators. In various

parts of the world, groups of workers are engaged in
special studies, not easily duplicable elsewhere. They
are expensive, they are a tremendous investment in terms
of time and staff, but they can be very rewarding scientifically. How can one get guidance as a public health
worker from such special research studies--from the experimental animal worker, from the expert on body composition, from the statistician, from the anthropologist,
etc.?

A great deal can be obtained. Not least, funding
may be available for a highly scientific study of inherent value, concerned, for example, with some specific
objective concerning mental development or the prevalence
of atheroma in later life. However, very useful additional information can be obtained from such a study concerning the validation and interpretation of anthropometric
indicators and indices, which the public health worker
needs for field work. These special studies are more
often longitudinal in nature, and, in this regard, there
comes to mind the vast amount of data already available
in different parts of the world at the present moment,
stored on tapes in different centers. This information
could be very useful, if analyzed.

Another matter that has come up repeatedly at this
meeting is the need to correlate anthropometric indicators
and indices with the severity, chronicity, and type of
PCM, and also, more fundamentally, with detailed function,
with prognosis, and with body composition, especially
stores of protein and fat.

There is a need to validate apparently simple measures, such as the chest-head ratio,[1] the "seven-year
height,"[5] and even the suggestion that the hand over the
head is a good screening "ratio" of whether the child is
five years or less. There is a particular need for guidance concerning <u>precise age-independent</u> anthropometry.[6]

<u>The design of apparatus</u>. Another form of guidance
or assistance that the field worker needs is in the design
of economical apparatus and its scientific testing in the
field. The cost-effectiveness of this type of intermedi-
ate technology is a paramount consideration. For example,

there is a need to devise a fatfold caliper that would
cost much less than $50.

   Evaluation of intervention programs. Guidance is
required in the evaluation of intervention programs,
including the need for revised regional growth standards
for young children.[7] While there is much evidence that
the weight and height of young children is more responsive to environment, and probably nutrition in particular,
than to genetic influences, it seems important to organize
up-to-date coordinated international anthropometric
studies, including especially measurements likely to be
valuable in the field.

CONCLUSIONS

   In conclusion, certain incorrect polar assumptions
are occasionally voiced. One is that nothing useful can
be done with simple anthropometric measures. This is not
true. Some such measures have been used successfully to
give imperfect, but useful, guidance. They are far from
perfect, and there is a great need for better measures
known to relate to body function and composition. At the
same time, practical action can be based on simple measures, provided it is realized that these approaches are
crude and approximate.

   The converse is the mirror-image delusion that the
interpretation of anthropometric data is simple. Plainly,
this is not the case, and great caution is needed.

   There are four main disciplines at this meeting--
the experimental animal nutrition worker, the statistician,
the physical anthropologist, the public health worker.
However, this by no means implies that we need have a
"quadrichotomy" of opinion. If one may use last year's
cliché, there is an enormous "interface" and overlap
between these different groups that can be immensely pro-
ductive, provided there is willingness to compromise.
It is necessary to use available approximations, provided
they can give information that is useful and worthwhile
from the public health viewpoint. At the same time, constant effort is needed to strive for the scientific

validation of these tests and to search for others, that can, in turn, be simplified and used in the field.

Plainly, the current need for this type of approach is urgent—not 20 years in the future, but <u>now</u>. There are nutritional problems all round the world, and guidance has to be given based on current knowledge.

For example, in India they are now setting up, or have set up, a national nutrition monitoring unit. Also, in Indochina, the Mekong Committee is planning some form of nutritional monitoring or surveillance for proposed post-war development schemes, which will mostly be concerned with water, with irrigation, with dams, with electrification, etc. Very wisely, it has been realized that they can upset the ecology, and produce a different life style for the communities involved. The Committee is trying to introduce a form of monitoring and surveillance to anticipate problems or to detect them early. Again, the methods need to be uncomplicated, inexpensive and based on the current state of the art.

In conclusion, a meeting like the present one, bringing together scientists and field workers, can be of immense value because it gives impetus to the application of science—to bridging the gap between the laboratory and Mankind.

REFERENCES

1. Jelliffe, D.B.: The assessment of the nutritional status of the community (with special reference to field surveys in developing regions of the world). WHO Monogr. Ser., 53:1, 1966.

2. Jelliffe, D.B. and Jelliffe, E.F.P.: The prevalence of protein-calorie malnutrition in Haitian preschool children. Amer. J. Publ. Hlth., 50:1355, 1960.

3. Pan American Health Organization: The National Food and Nutrition Survey of Barbados. Washington, 1972.

4.   Falkner, F.: Long term developmental studies: A
      critique. Early Devel., 51:412, 1973.

5.   Bengoa, J.M.: Significance of malnutrition and
      priorities for its prevention, p. 103. In:
      Berg, A., Scrimshaw, N.S. and Call, D.L. (Eds)
      Nutrition, National Development and Planning.
      Cambridge, MIT Press, 1973.

6.   Jelliffe, D.B. and Jelliffe, E.F.P.: Age-independent
      anthropometry. Amer. J. Clin. Nutr., 24:1377,
      1971.

7.   International Union of Nutritional Sciences,
      Commission III, Committee I, The Creation of
      Growth Standards: A Committee Report. Amer. J.
      Clin. Nutr., 25:218, 1972.

THE RELATION BETWEEN SIZE AT BIRTH AND PRESCHOOL CLINICAL SEVERE MALNUTRITION*

Joaquín Cravioto and Elsa R. DeLicardie

From the Scientific Research Division,
Hospital del Niño IMAN, México, D.F., Mexico

There has been a tendency to consider that, in situations in which food provision for infants and young children is marginal, intrinsically larger infants would presumably have higher nutritional requirements and are, therefore, at greater risk for the development of severe malnutrition.[1] From the public health point of view, an analysis of the relation between size at birth and the subsequent occurrence of severe malnutrition in early childhood is important. If size at birth helps to define the likelihood with which a child would develop severe malnutrition during infancy and the preschool years, such knowledge would be valuable in identifying children at risk and, potentially, for prevention.

In the course of a longitudinal study beginning at birth, 22 of the 334 infants being studied developed severe clinical malnutrition by four years of age. Such cases occurred despite the fact that all children in the cohort were examined by pediatricians biweekly, growth failures were identified, infectious and other illness

* Supported by Grants-in-aid from the Nutrition Foundation, Inc., The Foundation for Child Development (formerly Association for Aid of Crippled Children), The Van Ameringen Foundation, and The von Monell Foundation.

diagnosed and the parents given detailed advice on appropriate feeding and management.[2] The occurrence of severe
malnutrition under such circumstances makes possible
retrospective analysis of the etiology of severe malnutrition. Such an analysis obviously has many aspects, both
biological and social. In the present report, however,
we restrict our consideration to an examination of the
relationship between the size of the child at birth and
the occurrence of severe malnutrition during the first
four years of life.

SUBJECTS AND PROCEDURE

The overall design of our retrospective longitudinal
study has been described in detail elsewhere[2] and needs
only to be briefly summarized. The study is an ecologic
one in which a total cohort of 13 months of births (N=334)
is being followed from birth through the first school
years. Children are examined by pediatricians biweekly
and, at that time, they are weighed and measured. At
specified times, behavioral and familial evaluations are
made. The setting is a rural village of approximately
6,000 people in southwest Mexico. The data are gathered
through predefined protocols by a resident team of pediatricians, psychologists, social workers, nurses and
nutritionists.

Of all the children in the birth cohort, 78 percent
were born in their own homes under the care, in almost
all cases, of a trained semiprofessional local midwife.
These practitioners either did not weigh the child at
birth, or, if they did so, used uncalibrated scales, the
reliability and validity of which were highly questionable.
Therefore, it was necessary for birth weight to be obtained
by a pediatrician from the field team using recently calibrated equipment.

At the time of weighing, a physical examination of
the child was performed, anomalies were noted, and careful
measurements were made of total body length, head circumference, chest circumference, arm circumference, and

fatfold thickness. In the great majority of cases, these
observations were made on the day of birth; in most of
the remainder on the second day. In a few cases, first
weighings were delayed as much as 48 to 72 hours. Because
it was possible that these delayed measurements could
systematically decrease the magnitude of the birth weight
estimate, 100 children who had been weighed on the first
day of life were reweighed on the third day. No overall
systematic decrement was noted at reweighing; consequently
it was most unlikely that the small number of late weighings resulted in a systematic reduction in the estimation
of birth weight.

THE SEVERELY MALNOURISHED CHILDREN

As stated above 22 cases (14 girls and 8 boys) were
diagnosed as suffering from clinical severe malnutrition.
Age at the time of the diagnoses ranged from four to 53
months, with a single infant below one year of age, nine
cases between one and two years, eight cases between two
and three years of age, three patients with ages between
three and four years, and one case diagnosed at 53 months
of age. Fifteen of the 22 patients corresponded to the
kwashiorkor type, the other seven cases were of the marasmic variety. The proportion of marasmus in females and
males was 4:3, while the number of females with kwashiorkor
was twice the number for boys. Probably due to the small
number of cases, these differences are not statistically
significant at the 0.05 percent level of confidence.

RESULTS

As may be seen in Table I, mean weight, height, head
circumference, chest circumference, arm circumference and
fatfold thickness at birth are almost identical between
the index cases as a group and the cohort. Also the
variances of each of these measures are similar between
the groups and none of the differences between the means
is statistically significant.

Despite an absence of differences in growth attainment, it was possible that the interrelations among the

TABLE I

Size at Birth in Index Cases and Cohort as a Whole
(Land of the White Dust)

Measurement	Index cases		Cohort		t
	Mean	Standard deviation	Mean	Standard deviation	
Weight (g)	2855	417	2896	407	0.45
Height (cm)	48.3	1.9	48.0	2.1	0.63
Head circum- ference (cm)	33.9	1.0	33.7	1.4	0.61
Chest circum- ference (cm)	31.9	1.7	32.2	1.8	0.70
Arm circum- ference (cm)	9.9	1.1	10.0	0.9	0.44
Fatfold thickness (cm)	4.3	1.3	4.4	1.0	0.38

P is non-significant for each value of t.

growth measures at birth could differ between the two
groups. Consequently, the correlations among the growth
measures in the index group and the cohort were compared.
All the correlation coefficients were positive and significant indicating definite associations among these
measures in both groups (Table II). The groups did not
differ significantly in the probability levels of these
associations.

Since sex differences were possible, sex specific
comparisons were made between the index cases and the
cohorts both for mean values and for the associations
among measures. The findings for each sex were replicates of those for the whole group.

If size at birth were important in determining vul-
nerability to severe malnutrition, it would be most likely
to be reflected in the findings from those children who
develop malnutrition earlier in life. Therefore, from
this point of view, one would expect to find a significant relationship between growth achievement at birth
and the early development of severe malnutrition and,
it would be less likely that this relationship would be
found in children who become severely malnourished at
later ages. For this reason, we have explored birth
size in relation to the age at which severe clinical malnutrition occurred. The findings from this analysis
indicate no systematic relation between birth size and
age of incidence of severe clinical malnutrition.

DISCUSSION

Our findings indicate no systematic relationships
between size at birth and the development of severe clinical malnutrition in infancy and early childhood. Moreover, no systematic differences in size at birth are
associated with the age at which such severe malnutrition
occurs. At the practical level, such data suggest that
size at birth cannot be used as a good predictor of risk
for the subsequent development of severe clinical malnutrition. They lead to the hypothesis that the factors
that influence intrauterine growth may be quite different

from those that influence nutrition and growth in the postnatal period.

At the theoretical level, the findings provide no support either for the hypothesis that larger children are at greater risk for severe malnutrition or that small size at birth is due to the same processes that contribute to the development of severe malnutrition in infancy and early childhood. Thus, size at birth is unrelated to the occurrence of severe clinical malnutrition in the postnatal period.

TABLE   II

Intercorrelation Coefficients Among
Growth Measures in Index Cases
and Cohort

(Land of the White Dust)

	A	B	C	D	E	F
A	---	.54	.59	.86	.82	.65
B	.75	---	.49	.49	.37	.27
C	.71	.70	---	.61	.46	.32
D	.82	.63	.63	---	.76	.69
E	.75	.51	.52	.64	---	.78
F	.44	.29	.27	.40	.50	---

The coefficients that are underlined
refer to index cases.   The others
refer to the corhort.

A = weight; B = height; C = head
circumference; D = chest circumference;
E = arm circumference; F = fatfold.

It should not be assumed from these findings that perinatal conditions other than size are unrelated to the subsequent development of severe malnutrition. Behavioral attributes of the child, neurologic abnormalities, as well as peculiarities of the mother-infant interaction may all be relevant predictors of disturbed nutritional outcomes and the relationship of such factors to severe malnutrition will be considered in subsequent reports. However, the data of the present report indicate strongly the absence of a relationship between size at birth and severe clinical malnutrition in infancy and early childhood.

REFERENCES

1.   Garrow, J.S. and Pike, M.C.: The long-term prognosis of severe infantile malnutrition. Lancet, i:1, 1967.

2.   Cravioto, J., Birch, H.G., DeLicardie, E., Rosales, L. and Vega, L.: The ecology of growth and development in a Mexican preindustrial community. Report 1: Method and findings from birth to one month of age. Monogr. Soc. Res. Child Develop. Ser. No. 129, 34: No. 5, 1969.

ECOLOGICAL FACTORS RELATING TO CHILD GROWTH AND NUTRITIONAL STATUS

L. A. Malcolm

From the Department of Public Health
Konedobu, Papua New Guinea

In human ecosystems, the impact of the environment
is felt most forcibly on the growing child, particularly
in the early years of life. Exposure to a hostile world
after leaving the protection of the maternal uterus is
followed in all societies by a death rate which is reached
again only in old age. However, the range of environmental hazards to which the growing and surviving child
is exposed varies widely in different societies. Of all
the indices available to measure the impact of these
hazards, few are as informative or as sensitive as the
variation displayed in the physical growth of children.[1]

Growth is a dynamic process and the size and shape
of a child on the road from conception to the mature
adult is, at any stage, the resultant of competing, com-
plementary and interacting influences. The general course
and pattern of growth is laid down in a genetically programmed sequence at conception and it appears that only
strongly acting environmental influences can deflect
growth from this pre-determined course,[2] the main effect
being on the tempo rather than the pattern of growth.

Increase in size is dependent primarily on the supply
of sufficient building material of adequate quality. For
this reason, it could seem that the food intake of the
growing child is the main determining factor in ensuring

329

that programmed growth rates are achieved. However, the
quality and quantity of nutrients actually assimilated
and utilized by the growing organism are affected by a
wide range of influences deriving from the environment,
many of which interact with one another.[3] Nutrition is
only one of the factors of the ecosystem. This paper
examines various ecological characteristics that influence human growth and development, drawing particularly
on the Papua New Guinean situation for examples of these
influences. Three stages of growth are examined, prenatal,
preschool, and school age through to maturity, each stage
exhibiting distinctive features which require separate
treatment.

THE PRENATAL PERIOD

Growth begins at conception. Although birth is the
earliest stage at which growth can normally be measured,
significant influences have already led to wide variations[3] in size, as indicated by the range of mean birth
weights listed in Table I.[4-8] The heaviest mean weight,
reported by Barnes,[4] is of children whose parents were
living on the government station at Chimbu and were provided with rations, housing and health care. Almost as
heavy are children whose fathers are members of the Papua
New Guinea Defense Force and who are provided with a
reasonable income with family allowances, permanent housing, environmental protection and adequate health services.
The Lumi[7] and Inanwantan[8] people, on the other hand, whose
birth weights are lower than any others reported anywhere,
subsist on sago with little or no animal protein. Their
main protein source is green leaves. They are a primitive
group with almost no socioeconomic or cultural development
and with poor environmental and health services.

Fetal growth is influenced to only a minor extent
by the genotype as shown by the correlation between fetal
and maternal size but not paternal size.[2] Although the
fetus is an efficient parasite, and the intrauterine
environment offers a greater degree of protection than is
present postnatally, there is considerable evidence that
environmental influences are of paramount importance in
determining size at birth.[9,10] Factors related to the

intrauterine environment such as weight of the mother, parity, weight gain during pregnancy, as well as factors in the wider maternal environment such as tobacco smoking, malaria, nutrition and socioeconomic change have all been shown to influence prenatal growth. el-Oksh et al.[11] in a study of prenatal and postnatal influences on growth in mice, showed that the prenatal maternal environment accounts for 61 percent of the total variation in body weight at birth, while the fetal genotype was of no importance.

The specific role of nutrition in determining human birth weights has yet to be unequivocally demonstrated, but animal experiments, although not directly applicable to human growth due to the slower growing primate fetus, have shown that where protein malnutrition commenced prior to mating, both birth weight and brain size were significantly reduced compared with controls on an adequate protein intake.[12] Iyengar[13] demonstrated an

TABLE I

Mean Birth Weights of New
Guinean Infants from Some
Populations

Population	Weight (kg)
Chimbu station[4]	3.27
Lae army[5]	3.13
Chimbu village[4]	3.00
Rabaul[6]	2.96
Lae urban[5]	2.91
New Ireland[6]	2.64
Lumi[7]	2.40
Inanwantan[8]	2.32

increase in birth weights of infants born to Indian
mothers who received substantial protein and calorie
supplements in the last month of pregnancy.

Sex differences in birth weights suggest that some
genetic influence may be operating but even this has been
challenged recently by Ounsted and Ounsted,[14] who suggest
that the faster growth of the male may be due to the anti-
genic differences between the male fetus and its mother.
Further limiting factors on the growth of the fetus may
be related to the early biological experiences of the
mother. Ounsted and Ounsted[14] suggest that these limits
may be determined in part by the degree of constraint
imposed upon the mother when she herself was a fetus.

Recent work suggests that intrauterine infection may
influence fetal growth. Such infection has been indicated
by the reported presence of high levels of IgM in the
cord blood of Latin American new borns[15] whose mothers
were living in poor socioeconomic and sanitary conditions,
IgM not being transmitted across the placenta as is IgG.
These findings, confirmed in Papua New Guinean infants,
are important in view of the permanent effect on growth,
of malnutrition, infection and other influences early in
life, as demonstrated by animal experiments.[16,17]

It would seem reasonable to conclude, therefore,
that birth weight is determined to a very large extent
by the quality of the total environment prior to birth
rather than the genotype of either parents or fetus, and
that this early biological experience may permanently
influence the subsequent growth pattern of the child,
leaving an imprint on the ultimate characteristics of
the adult.

THE PRESCHOOL CHILD

The early extrauterine growth of children in all
preindustrial societies follows a typical course of rapid
growth for four to six months, followed by a flatening of
the growth curve. This pattern has been described by many
investigators.[18-22] The typical child in such societies

doubles its birth weight in three months, but trebling
of the birth weight does not occur until 15 to 18 months.
For comparison, the European infant doubles its birth
weight at five months and trebles its birth weight at
12 months. Figure 1 shows the range of variation of
growth in mean weights recorded in Papua New Guinean
societies.[5,21,22] The most rapid growth rates are

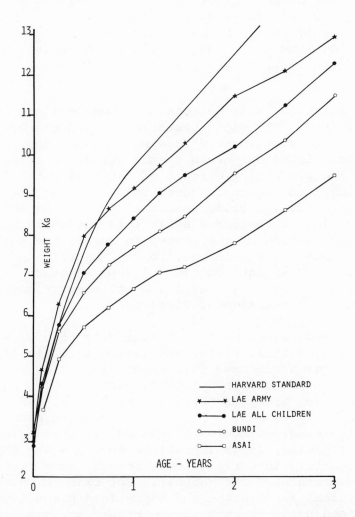

Figure 1. Weight for age of children from some Papua
New Guinean populations compared with Harvard standards.

achieved in urban children, particularly those of the
more privileged army community referred to above, whereas
the slowest rates are found amongst a primitive, poorly
contacted group subsisting on the tuberous sweet potato
as the basic staple with only three percent of the calorie
intake being derived from protein, and living in deprived
environmental and health circumstances. The curves
between these two extremes are from populations whose
environmental characteristics exhibit a corresponding
spectrum of variation. Associated and correlated with
this range of growth rates is a differential death rate
in children, especially in the one to four year old
child, varying from less than 4/1,000 per year in the
urban child[5] to 53/1,000 per year amongst the Asai
children.[22]

The reasons for this pattern have been extensively
reviewed.[18,20,21] The rapid postnatal growth can be
attributed to an unregimented feeding schedule and out-
standing lactational output in preindustrial societies,
the continuing protective effects of maternal antibodies,
and shielding from the environment for the first few
months of life through close association with the mother.
The interruption to the curve after this period is associated with the failure of the breast to supply the full
nutritional needs of the infant, to the insufficient
quality and delay in introduction of supplementary feeding
and to an increasing exposure to a hostile environment
favouring the transmission of disease.

Various ecological factors influencing this growth
pattern and nutritional status are listed in Table II,
classified after Scrimshaw[3] into headings of food availability, food needs and food intake, each of which has
elements deriving from the physical, biological and
socioeconomic environment. This list is not an exhaustive
one, and many headings could be expanded further. In any
such classification, it is impossible to segregate the
factors adequately into rigid compartments; many factors
interact and reinforce one another. For example, disease
impairs physical performance and hence food production,
increases food needs through tissue breakdown, and reduces
appetite and hence food intake.[23,24] In most societies,
socioeconomic patterns determining food intake are of

TABLE   II

Ecological Factors Influencing Nutritional Status

_____

Food availability

    physical -- land   -- terrain,   fertility

             -- climate   -- rainfall, drought,
                                        flooding & irrigation
                           -- temperature
                           -- food storage and
                                    contamination

    biological--agriculture -- types of food
                         -- insect and other pests

             -- storage   -- wastage, pests

socioeconomic --agriculture -- practice
                         -- extension efforts
                       --land tenure & utilization
                         -- new crops
                         -- credit

             -- education   -- schools
                         -- community development
                         -electronic communications

             -- food
               distribution-- storage
                         -- roads, transport
                         -- stores

             -- health   -- disease pattern
                            and control
                         -- health services

```
        --economic
           development--income, employment
                        --food costs and quality
                        --urbanization

        --political    --war,social oppression
                        --government policies

        --cultural
           attitudes --breast feeding
                      --family size, spacing
```

Food needs

```
    physical    --climate --heat, sweating losses
                          --cold

    biological             -- disease
                           -- age and rate of growth
                           -- pregnancy and lactation

    socioeconomic          -- activity and work load
```

Food intake

```
    physical               -- climate

    biological             -- disease
                           -- age and developmental
                                        status

    Socioeconomic  cultural -- ignorance, prejudice
                           -- traditional practices
                           -- maternal care
                           -- breast feeding practice

              food preparation
                           -- facilities, opportunity
                           -- motivation
                           -- consistency and bulk
                           -- meal frequency
```

much greater importance than food production in the causation of specific preschool malnutrition. Programs aimed at preventing malnutrition and improving child health must be based on what Jelliffe calls an ecological diagnosis,[24] i.e., an assessment of all the factors and their relative importance in the chain of multiple causation.

It is not intended to systematically discuss these factors further. This has been the subject of much previous effort. Rather, it is proposed to examine certain specific elements from the Papua New Guinean situation that appear to be of critical importance to the health of the preschool child.

In the past, much attention has focused on the non-availability of a suitable weaning diet as the major cause of malnutrition. Based on the assumption that this was due largely to the low protein content of the available food, considerable time, effort and money have been expended in developing, evaluating and promoting various protein rich mixtures to supplement the diet of the pre-school child.[25] These efforts, which have been repeated in many countries, including Papua New Guinea, have been rewarded with comparatively little success. Failure has been attributed to the cost of these supplements, the lack of discretionary income amongst the target groups to permit changes in food expenditure, the effort required in their local production and preparation,[26] and to the lack of recognition of these foods as being either culturally appropriate or necessary.

More recently the need for such protein supplements compared with the need for an increased total food intake has been questioned.[26-28] In many countries including Papua New Guinea, undernutrition and frank marasmus are far more common than is kwashiorkor. Furthermore, it has been shown that the correction of a deficient total food intake would at the same time reduce protein deficiency to an insignificant level. Few attempts, however, have been made to define the problem of food quality versus food quantity by anthropometric analysis of the vulnerable groups.

Of the different parameters that may be examined to determine the specific effect of certain nutrients, height is affected most by insufficient protein intake where energy intake is adequate, weight and arm circumference by both protein and energy intake, whereas skinfold thickness is a specific measure of calorie reserves and hence of total energy intake.[29]

Skinfold thickness studies (Figures 2 and 3) show that, compared with European children who reach a peak at or just beyond one year of age in both triceps and subscapular skinfold thickness,[30] children from both urban Lae and from a typical rural population of Kaiapit reach this peak prior to six months, following which there is a steady fall until the age of one and a half to two years. By this time a considerable gap has

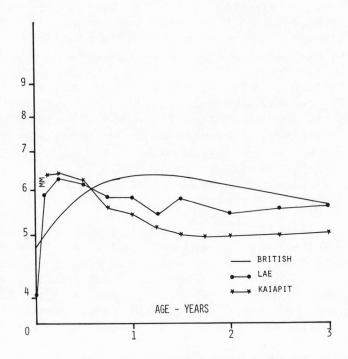

Figure 2. Subscapular skinfold thickness for Lae and Kaiapit children compared with British children.[30]

developed, especially in the triceps area, between
European and Papua New Guinean values, these latter being
less than 70 percent of the European figures at the age
of 15 to 18 months. Limited comparative data are available but a similar pattern at the same level of triceps
values was reported for Guatemalan children by Guzman
et al.,[19] while Neumann et al.[31] reported a peak at six
to nine months, but at a much lower value, for rural
Punjabi children. Deficits in other parameters are much
smaller proportionately. For example, the mean height
of the Lae urban child is 97 percent, weight 84 percent
and arm circumference 88 percent of the corresponding
mean values for European children at comparable ages.
Neumann et al.[31] showed that the triceps skinfold values
were similarly depressed in comparison with weight and
arm circumference figures.

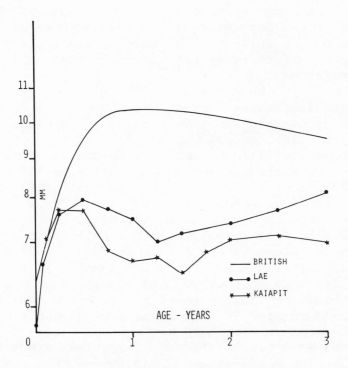

Figure 3. Triceps skinfold thickness for Lae and Kaiapit
children compared with British children.[30]

Although European norms may not be appropriate standards and absolute values may be affected to some extent
by climate, the figures and general patterns indicate
that total food intake is inadequate and is a major factor in determining impaired growth and consequently insufficient protein intake. It is not claimed, of course,
that improved food intake of whatever quality will solve
all nutritional problems in the preschool child. Bulky
food and low protein value may still be important.[26]
However, the feeding of sufficient quantity of whatever
food is available is the first step in attacking the
problem. The addition of quality food, with all the
implications this involves, is the second step to be
embarked upon only after the first has been taken.

The second ecological factor of some importance
affecting growth is the cultural attitude in a society
to child care and feeding. Few observers appear to have
asked the obvious but perhaps important questions, "Why
does a mother feed her child?", "What determines whether
a mother feeds her child at a particular time?", and
"What motivating factors operate in child feeding in
different cultures?"

The Western observer, examining the socioeconomic
patterns of people in primitive cultures, too often
interprets his findings, including child care patterns,
using concepts familiar to himself from his own cultural
background. Normally the Western mother feeds her child
to satisfy her own preconceived need of the child. In
many situations this concept of need is in excess of the
actual need. Feeding is based on a timetable and success
measured by satisfactory weight gain. Failure to attain
certain predetermined goals, either in food intake or
weight gain, may lead to distress and anxiety in the
Western mother. To the mother in a primitive culture,
factors determining feeding practice may be quite different. Child rearing in most Papua New Guinean societies
is permissive and unrestricted and much of a child's
activity such as sleeping, eating, excreting, etc., is
determined by the personal choice of the child.

Studies of the mother-child relationship using prolonged, unobtrusive and unobstructed film sequences

(Sorenson, personal communication) suggest that child feeding in Papua New Guinea is based almost entirely, and at all ages, even from birth, on the demands of the child. It is the child who determines when and with what he is fed. In the early weeks of life, the breast is immediately available when he cries and is given without hesitation at any time of the day or night. As soon as the infant is old enough to grasp the breast, he will take the nipple himself. Thus the infant is capable of satisfying his own needs, by his own demands, with the breast until the age of six months. However, beyond this time or in the event of lactational failure, his own demands for the food resources immediately available to him are insufficient to maintain an adequate nutritional intake. It is only later, in the second year of life, that he becomes sufficiently mobile and vocal to satisfy his needs through his own efforts. Therefore, a gap exists between the age of six months and one and a half to two years during which this demand/permissive type of feeding regime fails to satisfy the nutritional requirements of the child.

Mothers appear reluctant to actively encourage their children to eat food at any age. Frequently it is noted that attempts in hospital to feed a sick or lethargic child may distress the mother, her attitude being that, because the child does not want to be fed, he should not be compelled to take food.

These findings suggest that the quantitative deficit described above may be largely due to maternal failure to appreciate this demand gap and to mobilize her own efforts to fill the gap from whatever food might be available. Furthermore, if demand, rather than need, is the main motivating element in child feeding, a nutrition education program which presumes a developed concept of need will make little impact, particularly if based on the introduction of culturally foreign foods. An awareness of the child's needs must be implanted in the mother before she will show any concern for either the quantitative or qualitative introduction of supplementary food.

The importance of disease as a determinant of child health and growth has been extensively discussed by many

writers[2,3,7,10,15,16,19,23,24,32] and high morbidity, as
well as mortality, rates appear to be associated with
the variations in growth rates of the preschool child,
as described above. It is well established that suboptimal nutrition impairs resistance to infection and
that infection in turn impairs nutritional status. The
impact of infection per se on growth, however, remains
uncertain as does the link between growth rates and the
prevalence of overt clinical protein-calorie malnutrition.

While there is no doubt that illness impairs growth
temporarily, this may be succeeded by a catch-up phase,
on recovery, that restores body weight to the channel
being followed prior to the infection. Guzman et al.[19]
found no relationship between physical growth and the
nature and amount of illness or between days of illness
and growth in either height or weight in Guatemalan
children. No differences in growth rates were found by
Morley et al.[33] between children protected from malaria
by antimalarials compared with a control group and, while
malaria prophylaxis during and after pregnancy resulted
in larger babies, this difference in weight was lost during the first few months of life. McGregor et al.[34]
found little difference in the size and rates of growth
in Gambian children who died compared with surviving
children and, although there was a marked seasonal variation in growth rates, this could not be clearly identified with a higher prevalence of infection.

Recent studies of Bundi people,[21] a group of some
6,000 to 7,000 New Guinean highlanders, suggest that
there is little relationship between growth rates on the
one hand and the prevalence of overt clinical protein-
calorie malnutrition and preschool death rates on the
other. Prior to 1959, when significant contact with
these people commenced and health and medical services
were provided, the prevalence of nutritional disease
requiring active treatment was high, being about three
percent of preschool children. Kwashiorkor was seen as
frequently as marasmus. At this time, the one to four
year death rate was estimated from census data and
Catholic Mission records to be in excess of 50/thousand
children in this age group per year. By 1967, over a
period of seven years, the prevalence of clinical

malnutrition, now predominantly marasmus, had fallen to
less than 0.5 percent of preschool children while the one
to four year death rate had fallen to about 10/thousand.

Despite these dramatic changes, child health records
of weights showed no significant increase. The reasons
for this decline in disease and death rates are far from
clear. While considerable educational, political and
social changes occurred over this period, there was virtually no economic development to permit changes in traditional food consumption and the importation of protein-
rich foods such as milk, tinned fish or meat or even rice.
However, the latter factors are considered to be of major
importance in the declining prevalence of clinical malnutrition and lowered toddler mortality rates elsewhere
in Papua New Guinea. Only one trade store existed in
this whole population in 1967 and its sales of protein-
rich foods were insignificant.

Three factors that may have been of significance,
associated with the introduction of health services, were
the nutrition education program of the child health service in which mothers were encouraged to introduce supplementary feeding within the first few months, the immuni-
zation program against pertussis and the provision of a
simple treatment service for common illnesses such as
pneumonia, diarrhea, malaria and skin disease at aid-
posts. If a significant quantitative change had occurred
in the supplementary feeding pattern, some increase in
weight for age might have been anticipated. Because no
change occurred and because clinical malnutrition or
alternatively death may be precipitated rapidly by a disease episode in a child with marginal nutritional status,
the changing prevalence of disease and death could be
largely attributable to the provision of a simple preventive and treatment service.

A low prevalence, one percent of clinical malnutrition, entirely marasmus, was found also amongst Asai preschool children[22] whose growth rates are lower than those
of any other reported population. The toddler death rate
has been high--53/1,000 but no information was available
on the rates current at the time of this study. Here
again, a simple treatment and preventive service had been

provided that may have been an important factor in the
low prevalence of overt malnutrition.

These findings, as well as those reported above,
tend to suggest that growth per se is not a good index
of nutritional disease nor of the impact of a health
program. Furthermore, simple and inexpensive health
programs may greatly ameliorate the harmful consequences
of a poor nutritional environment without any corresponding change in nutritional intake or in rates of growth.
A nutrition improvement program that ignores the benefits
of disease control and seeks to apply a purely nutritional
solution, to what is seen to be a purely nutritional problem, is a tunnel vision approach that fails to appreciate
the much wider perspective of a deprived environment.
Malnutrition is largely a social problem and its solution
should be sought within the total context of the social
and socioeconomic characteristics that make up the complex factors in the ecosystem.

SCHOOL AGE AND ADOLESCENCE THROUGH TO MATURITY

Beyond the preschool years, growth curves of children
in different Papua New Guinean populations continue to
diverge, as shown in Figure 4, compared with European
children.[35] These curves are based on measurements of
children of known age[5,21] with the exception of the Asai,[36]
where dental eruption was used to locate critical points
on the graph, eruption times being only slightly and
predictably delayed by slower growth. The different rates
of growth are maintained throughout the growth span and,
although slow growing children grow for a longer period,
a substantial height deficit remains in these groups when
growth eventually ceases. Maximum divergence in height
occurs during the period of the adolescent spurt, which
commences in all groups at around nine years in girls
and 11 years in boys, but is much smaller in velocity in
slowly growing children. As a consequence, Bundi and
Asai children reach the height of European children of
10, 16, and 18 years respectively. On the other hand,
the urban child is only marginally shorter in height than
the European child of the same age, being one year of
height age behind at the point of maximum divergence.

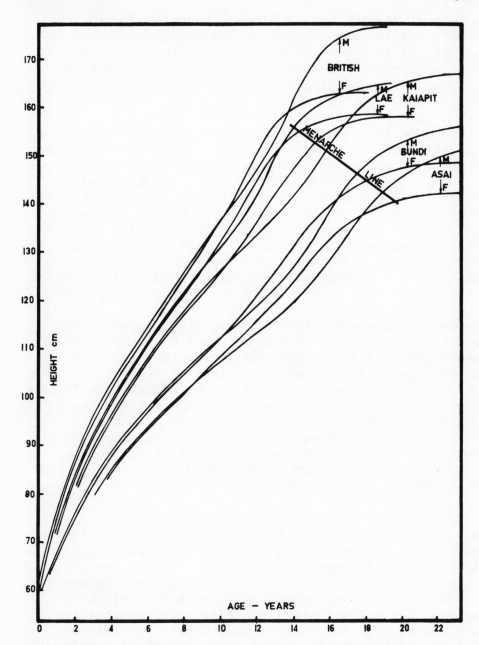

Figure 4. Height for age of children from some Papua New Guinean populations compared with British children.[35] M = male; F = female.

Sexual development is closely correlated with growth rates,[21] the "menarche line" on the graph indicating at its point of intersection with the female curve, the mean age of menarche in that population. This varies from 13.8 years in urban girls[5] to around 18.0 years in Bundi girls.[21]

More detailed studies of this growth pattern suggest that, although the overall pattern, such as the sex difference, the adolescent spurt and the general shape of the curves, is genetically determined, the differences between populations are almost entirely the product of the environment. The evidence on which this conclusion is based will now be considered. Urban children, whose genetic background is considered to not differ significantly from rural children, display rates of growth almost equal to those of European children,[5] as a consequence of their being reared in the better urban environment.[5]

In a longitudinal growth study,[21] children in the Bundi Catholic Mission boarding school environment were 4 cm shorter and 2.5 kg lighter in weight than their village counterparts after four years of residence. After eight years, they had recovered to the normal size of the village child of the same age. The same children, when fed supplementary protein in the form of skim milk powder, have, over a four-year study period, shown rates of growth equal to those of European children of the same age, and hence are 7.3 cm taller and 3.1 kg heavier than the equivalent village child.[29]

Bundi children reared in the village displayed a small but significant variation in growth rates with their position within the family.[21] While the first child is nearly 1 cm taller than the mean, with increasing position in the family there is a decline in height and weight, so that the third and fourth children are nearly 1 cm shorter. Thereafter, size increases again so that the sixth and subsequent children are larger even than the first child. A possible explanation of what is obviously an environmental variation, is that the first child is born to a fit and usually well-nourished mother. Subsequent children have a lesser advantage in this respect and successively compete for available food. Children born later, however, have the advantage of care, including feeding, by their older siblings, who may also contribute to the gathering of the family food and may forage on behalf of their younger siblings for titbits of animal protein such as insects, frogs and rodents.

Correlations between parent and child in both height and weight, normally of the order of 0.45 to 0.50 in European families, are depressed significantly in Bundi children,[21] an indication that this genetic relationship is being blurred by environmental pressures here as in Mexico.[37] One factor is a loss of weight in women due to an increased work load associated with marriage.

All the evidence suggests that the only environmental factor of any real significance at this stage, particularly in determining the variation in growth in height, is the level of protein intake. Supplementary feeding experiments in Bundi children[21] show that height gains were highly significant when protein was added to the diet, but no change was observed when only calories were added in the form of margarine. Height gains, among a wide range of anthropometric parameters, gave the best measure of discrimination between different levels of protein intake.[29]

The growth in height of the urban child appears to be related to the higher protein content of his diet[5]-- 12 percent of the calories being derived from protein, compared with only three to five percent in the diet of the village child.[21] Studies of the urinary output of nitrogen indicate that the protein intake of the urban child is in excess of FAO/WHO requirements.[5]

The effect of illness on the growth pattern at this stage, as for the preschool child, appears to be relatively insignificant. On the other hand, the addition of protein to the diet had a highly significant effect in diminishing the rate of illness in Bundi school children.[29]

CONCLUSIONS

Ecological factors appear to be of major importance in determining size at any age from conception through to maturity. While nutritional intake at any stage is of paramount importance, a wide range of factors including infection may also be significant in the prenatal period.

In the preschool child, quantity of food intake is the major determinant of size, particularly in the six months to two year period, while food quality, particularly the level of protein, is of secondary importance. Other factors including infection are significant only to the extent that they influence the actual level of food intake. Of these, the quality of maternal care, particularly maternal appreciation of the child's needs, is a major factor determining food intake.

There appears to be only a tenuous connection between the rates of growth in preschool children on the one hand and the prevalence of clinical protein-calorie malnutrition and toddler death rates on the other, the latter possibly being related more to the quantity of disease and its control in a community than to anthropometric nutritional status.

The rate of growth of the school age child is closely related to the level of protein intake with calorie levels being important only in relation to total food intake. Protein deprivation throughout the growth span results in a reduction of ultimate adult stature to an extent proportional to the degree and duration of the deprivation. Growth in height is the best anthropometric index of discrimination between different levels of protein inadequacy.

The solution to the problem of malnutrition must be sought, not within the narrow confines of a nutrition program, but through an examination of the complex of factors that make up the characteristics of the whole ecosystem, with an endeavour to influence those elements that are both susceptible to change and that constitute major stumbling blocks in the chain of causation of nutritional disease.

REFERENCES

1.   Tanner, J.M.: The Biology of Human Adaptability
     from Growth & Physique in Different Populations
     of Mankind. Clarendon, Texas, Clarendon Press,
     1966.

2.   Tanner, J.M.: Growth at Adolescence, 2nd Ed.,
     Oxford, Blackwell Sci. Publ., 1962.

3.   Scrimshaw, N.S.: Ecological factors in nutritional
     disease. Amer. J. Clin. Nutr., 14:112, 1964.

4.   Barnes, R.: A comparison of growth curves of infants
     from two weeks to twenty months in various areas
     of the Chimbu sub-district of the eastern highlands
     of New Guinea. Med. J. Aust., ii:262, 1963.

5.   Malcolm, L.A.: Health and nutrition, growth and
     development of children in urban Lae, Papua New
     Guinea. In preparation.

6.   Scragg, R.F.R.: Birth weight, prematurity and growth
     rate to thirty months of the New Guinea native
     child. Med. J. Aust., i:128, 1955.

7.   Wark, L. and Malcolm, L.A.: Growth and development
     of the Lumi child in the Sepik district of New
     Guinea. Med. J. Aust., ii:129, 1969.

8.   Oomen, H.A.P.C.: The Papuan child as a survivor.
     J. Trop. Pediat., 6:103, 1961.

9.   Bergner, L. and Susser, M.: Low birth weight and
     prenatal nutrition: An interpretative review.
     Pediatrics, 46:946, 1970.

10. Pachauri, S., Marwah, S.M. and Rao, N.S.N.: A
     multifactorial approach to the study of the factors
     influencing birth weight in the urban community
     of New Delhi. Indian J. Med. Res., 59:1318, 1971.

11. el-Oksh, H.A., Sutherland, T.M. and Williams, J.S.:
    Prenatal and postnatal maternal influence on
    growth in mice. Genetics, 57:79, 1967.

12. Chase, H.P., Dabiere, C.S., Welch, N.N. and O'Brien,
    D.: Intra-uterine undernutrition and brain
    development. Pediatrics, 47:491, 1971.

13. Iyengar, L.: Effects of dietary supplements late
    in pregnancy on the expectant mother and her
    newborn. Indian J. Med. Res., 55:85, 1967.

14. Ounsted, C. and Ounsted, M.: Effect of Y chromosome
    on fetal growth-rate. Lancet, ii:857, 1970.

15. Lechtig, A. and Mata, L.J.: Levels of IgG, IgA and
    IgM in cord blood of Latin American newborns from
    different ecosystems. Rev. Lat. Amer. Microbiol.,
    13:173, 1971.

16. Dubos, R., Savage, D. and Schaedler, R.: Biological
    Freudianism: Lasting effects of early environ-
    mental influences. Pediatrics, 38:789, 1966.

17. McCance, R.A.: The effect of calorie deficiencies
    and protein deficiencies on final weight and
    stature. In: Calorie Deficiencies and Protein
    Deficiencies. London, Churchill, 1967.

18. Jelliffe, D.B.: Infant nutrition in the sub-tropics
    and the tropics. WHO Monogr. Ser., 29:1, 1968.

19. Guzman, M.A., Scrimshaw, N.S., Bruch, H.A. and
    Gordon, J.E.: Nutrition and infection field
    study in Guatemalan villages, 1959-1964.
    VII. Physical growth and development of preschool
    children. Arch. Environ. Health, 17:107, 1968.

20. Bailey, K.V.: Growth of Chimbu infants in the New
    Guinea Highlands. J. Trop. Pediat., 10:3, 1964.

21. Malcolm, L.A.: Growth and development in New
    Guinea: A study of the Bundi people of the Madang
    district. Inst. Hum. Biol., Madang, Monogr.
    Ser. 1, 1970.

22. Malcolm, L.A.: Growth, malnutrition, and mortality
    of the infant and toddler of the Asai valley of
    the New Guinea highlands. Amer. J. Clin. Nutr.,
    23:1090, 1970.

23. Scrimshaw, N.S., Taylor, C.E. and Gordon, J.E.:
    Interactions of nutrition and infection. WHO
    Monogr. Ser., 57, 1968.

24. Jelliffe, D.B.: The assessment of the nutritional
    status of the community (with special reference
    to field surveys in developing regions of the
    world). WHO Monogr. Ser., 53, 1966.

25. Cordaro, J.B. and Call, D.: Nutritional protection
    of vulnerable groups through protein-rich mixtures:
    A critical review. Proc. IXth Internat. Congr.
    Nutrition, Mexico City, 1972.

26. Sukhatme, P.V.: Incidence of protein deficiency
    in relation to different diets in India. Brit.
    J. Nutr., 24:477, 1970.

27. Gopalan, C.: Kwashiorkor and marasmus: Evolution
    and distinguishing features. In: Calorie
    Deficiencies and Protein Deficiencies. London,
    Churchill, 1967.

28. Jelliffe, D.B.: Approaches to village-level infant
    feeding. The essential characteristics of weaning
    foods. J. Trop. Pediat., 17:171, 1971.

29. Malcolm, L.A.: Anthropometric, biochemical and
    immunological effects of protein supplements in
    a New Guinean highland boarding school. IXth
    Internat. Congr. Nutrition, Mexico City, 1972.

30. Tanner, J.M. and Whitehouse, R.H.: Standards for
     subcutaneous fat in British children. Percentiles
     for thickness of skinfolds over triceps and below
     scapula. Brit. Med. J., i:446, 1962.

31. Neumann, C.G., Shanker, H. and Uberoi, I.S.:
     Nutritional and anthropometric profile of young
     rural Punjabi children. Indian J. Med. Res.,
     57:1122, 1969.

32. Scrimshaw, N.S., Guzman, M.A., Flores, M. and
     Gordon, J.E.: Nutrition and infection field study
     in Guatemalan villages, 1959-1964. V. Disease
     incidence among preschool children under natural
     village conditions, with improved diet and with
     medical and public health services. Arch.
     Environ. Health, 16:223, 1968.

33. Morley, D., Bicknell, J. and Woodland, M.: Factors
     influencing the growth and nutritional status of
     infants and young children in a Nigerian village.
     Trans. Roy. Soc. Trop. Med. Hyg., 62:164, 1968.

34. McGregor, I.A., Rahman, A.K., Thompson, B.,
     Billewicz, W.Z. and Thomson, A.M.: The growth of
     young children in a Gambian village. Trans. Roy.
     Soc. Trop. Med. Hyg., 62:341, 1968.

35. Tanner, J.M., Whitehouse, R.H. and Takaishi, M.:
     Standards from birth to maturity for height,
     weight, height velocity and weight velocity:
     British children, 1965. Arch. Dis. Childh.,
     41:613, 1966.

36. Malcolm, L.A.: Growth of the Asai child of the
     Madang district of New Guinea. J. Biosoc. Sci.,
     2:213, 1970.

37. Mariscal, C., Viniegra, A. and Ramos Galván, R.:
     Predicción de talla en niños desnutridos y talla
     de sus progenitores. Bol. Med. Hosp. Inf. Mexico,
     23:465, 1966.

CONFERENCE PARTICIPANTS

Brožek, Josef M., B.A., Ph.D., Professor, Institute
of Psychology, Lehigh University, Bethlehem,
Pennsylvania 18015 (USA).

Buzina, Ratko, M.D., Sc.D., D.P.H., Department of
Nutrition, Institute of Public Health,
Rockefellerova 7, Zagreb (Yugoslavia).

Cheek, Donald B., M.D., D.Sc., Director, Research
Foundation, Royal Children's Hospital, Flemington
Road, Parkville, Victoria 3052 (Australia).

*Chow, Bacon F., Ph.D., Professor of Biochemistry,
The Johns Hopkins University, School of Hygiene
and Public Health, 615 North Wolfe Street,
Baltimore, Maryland 21205 (USA).

Cravioto, Joaquín, M.D., Professor, Chairman, Division
de Investigación Cientifica, IMAN, Hospital
Infantil IMAN, Insurgentes Sur 3700, México 22,
D.F. (México).

Falkner, Frank, M.D., F.R.C.P., Director, The Fels
Research Institute, Yellow Springs, Ohio 45387 (USA).

Goldstein, Harvey, B.Sc., National Children's Bureau,
One Fitzroy Square, London, W1P 5AH (England).

Habicht, Jean-Pierre, M.D., Ph.D., M.P.H., Head,
Biomedical and Epidemiological Section, Division of
Human Development, Instituto de Nutrición de Centro
América y Panamá, Apartado Postal 11-88, Guatemala
City, Guatemala (Central America).

Jelliffe, Derrick B., M.D., F.R.C.P., D.C.H., Head,
Division of Population, Family and International
Health, University of California, School of Public
Health, Los Angeles, California 90024 (USA).

* Recently deceased.

Jelliffe, E. F. Patrice, R.N., M.P.H., School of Public
    Health, University of California, Los Angeles,
    California 90024 (USA).

Johnston, Francis E., Ph.D., Professor, Department of
    Anthropology, University of Pennsylvania, University
    Museum, 33rd and Spruce Streets, Philadelphia,
    Pennsylvania 19104 (USA).

Malcolm, Laurence, A., M.D., F.R.C.P. (Edinburgh)
    Health Planning Unit, Department of Public Health,
    P.O. Box 2084, Konedobu, Papua (New Guinea).

Metcoff, Jack, M.S., M.D., M.P.H., Professor of
    Pediatrics and Professor of Biochemistry and
    Molecular Biology, The University of Oklahoma
    Health Sciences Center, P.O. Box 26901, Oklahoma
    City, Oklahoma 73190 (USA).

Pařízková, Jana, M.D., Ph.D., C.Sc., Head, Výzkumný
    Ústav Tělovýchovny, F.T.V.S., Charles University,
    Újezd 450, Praha 1 (Czechoslovakia).

Roche, Alexander F., M.D., Ph.D., D.Sc., Senior
    Scientist, The Fels Research Institute, Yellow
    Springs, Ohio 45387 (USA).

Terada, Harumi, M.D., Professor of Anatomy, Kitasato
    University School of Medicine, 1 Asamizodai,
    Sagamihara-shi, Kanagawa-ken, 228 (Japan).

Widdowson, Elsie M., D.Sc., Ph.D., Dunn Nutritional
    Laboratory, Milton Road, Cambridge CB 4 1XJ
    (England).

Wolánski, Napoleon L., Ph.D., Head, Department of Human
    Ecology, Polish Academy of Sciences, ul. Nowy Świat
    72, 00-330 Warszawa (Poland).

Adaptation, 230
Adenylic kinase, 95-8, 100-5, 107
Adipose tissue, 123-4, 128, 133, 150, 154, 161
————, and metabolic activity, 124, 130
————, cell number, 55, 62-3, 120, 130, 176-7
————, cell size, 55
————, changes in rehabilitation, 176
————, collagen, 62
————, estimate from height, 60
————, fatfolds, see separate listing
————, in maternal diabetes, 49-50, 52, 55
————, in muscle, 50
————, in obesity, 236
————, in pubescence, 56-7
————, in senescence, 125
————, metabolism of, in muscle, 125
————, response to insulin, 57, 60, 64
Adrenal gland, 49, 53, 60
Aerobic capacity, 130, 133, 136-7, 141
Age classes, 257-8
Age-independent anthropometry, 7-8, 159, 273
Altitude, high, 82
Amino acids in plasma, 3
Anthropometry, abdominal circumference, 273, 280
————, age-independent, 7-8, 159, 273
————, and nutritional status, 4, 15-25, 150, 220, 286,
    338
————, arm circumference, 4-5, 7, 159, 273, 280
————, arm circumference/ arm length, 7
————, arm circumference/ height, 5-7, 160
————, arm muscle circumference, 275, 280, 285
————, biacromial diameter, 275, 280
————, biacromial/bicristal diameter, 272
————, bicristal diameter, 275, 280
————, calf circumference, 275, 280
————, chest circumference, 275, 280
————, chest circumference/head circumference, 7, 317
————, fat area, 5
————, fatfolds, see separate listing
————, head circumference, 24
————, height, see separate listing

Anthropometry, laterality/linearity, 272
————, muscle area, 5
————, Quetelet index, 7, 20-2, 28, 240
————, reliability, 16-8
————, selection, 5
————, sensitivity, 16, 21-5
————, sitting height/standing height, 272
————, skinfolds, see fatfolds
————, specificity, 16-8, 20, 24-5, 286
————, stature, see height
————, weight, see separate listing
————, weight/height, 7, 20-2, 28, 240
————, weight/height$^{1.6}$, 8
————, weight/height^2, 270-1
Apparatus, basis for selection, 6-7
————, design, 317-8
————, scales, 6-7, 18
Appetite, and exercise, 48

Basal metabolism, 28
Biological status, 246, 249, 265
Birth order, 4, 227, 258, 346-7
Birth size, and diet during pregnancy, 122
————, and litter size, 122
Birth weight, 55
————, and intrauterine infection, 332
————, and later malnutrition, 320-7
————, and protein intake, 330-1
————, genetic factors, 330-1
————, in maternal diabetes, 50, 121-2
————, maternal factors, 197-201, 331-2
————, reference standards, 228
————, variability, 330
Birth, weight gain after, 55
Blood vitamin levels, 3
Body build, and exercise, 122, 133, 136, 138-9
————, urban-rural differences, 282
Body composition, 16, 27, 47, 57, 150-62, 164-79, 182
————, adipose tissue, see separate listing
————, aerobic capacity, 130
————, and body shape, 132
————, and body size, 132
————, and edema, 157-8, 161

Body composition, and exercise, 131, 133, 136, 138-9, 14
———, and functional capacity, 132
———, and litter size, 123
———, and maternal diet, 122
———, and nutritional status, 317
———, and rate of maturation, 132
———, extracellular space, 53, 157
———, in maternal diabetes, 49-55
———, in obesity, 47, 56-7, 60, 64, 236
———, intracellular mass, 57
———, lean body mass, see separate listing
———, muscle, see separate listing
———, prenatal, 165
———, thiocyanate space, 157
———, total body water, 158
Body density, 27
Body fat, see adipose tissue
Body size, in malnutrition, 169, 184, 186-9
———, urban-rural differences, 251, 259
Body surface area, anthropometric estimation, 28, 40
———, methods of measuring, 28, 40
Body volume, anthropometric estimation, 28
———, methods of measuring, 28, 40
Bone age, see skeletal age
Brain growth, age of cell division, 81
———, and growth hormone, 54-5
———, and insulin, 54,64
———, and maturation in diabetes, 64
———, and size at birth, 331
———, sparing, 167-8, 171
Brain maturation, 55, 168, 171
Breast feeding, 48, 56, 64

Calories, 1-5
Caloric intake in infancy, 55-6
Catch-up growth, 342
Cell number in malnutrition, 168, 172, 178
Cell size in malnutrition, 168
Cerebrum, cell number and size, 50
———, water content, 52
———, zinc content, 55
Child feeding, 340-1
Child rearing, 340-1

Circumferences, see anthropometric measures
Climate, 232
Collagen in adipose tissue, 62
Conditioned emotional response, 186-9
Congenital anomalies, 81
Contourograph, moiré, 30-40
Creatinine nitrogen, 3, 8
Critical period, 85
Cross-sectional studies, 287-99, 314-5
Cyrtographometer, 29-30

DNA synthesis, 105-7
Density, see body density
Developmental age, 239, 242
Diameters, see anthropometric measures
Diabetes, see maternal diabetes
————, sensitivity of pancreas, 64
Diet during pregnancy, see maternal diet
Disease, 4, 24, 246, 334, 342-3, 347
————, and prenatal growth, 82
————, intrauterine infection, 332
Diurnal variation, 260-1
Dwarfism, pituitary, 54

Ear size, 174
Edema, 21, 157-8, 161
Energy turnover, 123-4, 129
Environmental factors, sensitivity to, 230, 238, 240-1
Enzymes, see leukocytes
Exercise, 123, 127-8, 139, 249
————, aerobic capacity, 130, 133, 136-7, 141
————, and appetite, 48
————, and body build, 122, 133, 136, 138-9
————, and body composition, 64, 123, 131, 133, 136, 138-9, 142
————, and longevity, 48
————, and malnutrition, 132
————, and muscle growth, 48
————, and myocardial lesions, 131
————, and nutritional status, 232
————, and obesity, 236, 238
————, change with age, 129
————, in pubescence, 123, 139

Exercise, in senescence, 123, 139
Extracellular volume, 53, 157

Fat, see adipose tissue
Fatfolds, 5, 9, 28, 162, 280, 286, 338-9
———, in maternal diabetes, 121-2, 132-3, 280, 338-9
———, in newborn, 132
———, in prematurity, 132
———, suprailiac, 132
———, triceps, 5, 9
Fetal death, 73, 75, 77
Fetal malnutrition, 73-80, 82-4, 164-8, 184, 186-9, 202-11
———, DNA and RNA synthesis in maternal leukocytes, 106-8
———, and congenital anomalies, 81
———, and functional changes, 81-2
———, and respiratory distress syndrome, 81
———, cell replication, 74
———, fetal head size in, 108
———, hypoglycemia in, 81
———, maternal leukocyte enzymes in, 95-8, 100-5, 107
———, organ changes in, 80-1
———, synapses in, 81
———, types, 85-8
Food intake, and hormone secretion, 48
———, total, 337, 340
Food utilization, in malnutrition, 131-2
Formula feeding, 56
Functional capacity and rate of maturation, 131-2

Genetic factors, and growth, 9, 330
Grandmother's nutrition, 85
Gross national product, and height, 239
Growth, and birth order, 4
———, and disease, 342-3, 347
———, and litter size, 122
———, and seasons, 342
———, and socioeconomic status, 4, 9, 11
———, genetic factors, 9
———, prenatal, 330
———, retarded in utero, see fetal malnutrition
Growth charts, see reference standards

Growth hormone, 48-9, 54-5, 60, 191-2
Growth rate, 262

Hair root, 4
Head size, 174
Height, 5, 19, 23-4
————, and malnutrition, 282, 332-4, 342-3, 347
————, and gross national product, 239
————, and obesity, 56
————, and protein intake, 240, 334, 338, 347
————, changes in, 23-4
————, preschool pattern, 332-4
————, reference standards, 229
Hematocrit, 53, 252
Hemoglobin, 235, 252
Heterosis, 236, 238, 243-4
Heterozygosity, 236-8
High altitude, 82
Hormones, see individual glands
Hormone secretion and food intake, 48
Hydroxyproline excretion, 3
Hypoglycemia in fetal malnutrition, 81
Hypothalamus, 54

Ilial bypass, 60, 64
Illness, see disease
Infancy, caloric intake in, 55-6
————, obesity in, 2, 10-1
————, mortality, 243-4, 334, 343
Infection, birth weight and, 332
Insulin, 47-9, 54, 64
————, and brain growth, 54, 64
————, muscle response to, 57, 60
————, response of adipose tissue to, 57, 60, 64
Intracellular mass, 57
Iron in serum, 235

Kwashiorkor, 21, 323, 342-3

Lactation, and protein supplementation, 195-6
————, in preindustrial societies, 334
Lean body mass, 49-50, 53, 56-7, 64, 236
Leukocyte enzymes, 95-8, 100-5, 107

Litter size, and body composition, 123
————, and growth, 122
Longevity, and exercise, 48
————, and obesity, 64
Longitudinal studies, 288–90, 298–301, 312–4
Low–birth–weight babies, 74, 76, 79, 82
Lung volume, residual, 42
Lysine, 204–13, 215

Marasmus, 323, 342–3
Malnutrition, 2–4, 16–17, 21, 159, 169, 184, 186–9
————, and exercise, 132
————, and growth hormone, 191–2
————, and infant mortality, 334, 343
————, and menarche, 346
————, and postnatal growth, 169
————, and protein supplementation, 20, 154, 184, 211–2, 337, 343, 346–7
————, and rate of maturation, 252, 328, 346
————, and sexual maturation, 178
————, body composition in, 16–7
————, body size in, 235
————, brain maturation in, 168, 171
————, brain sparing in, 167–8, 171
————, cell number in, 168, 172, 178
————, cell size in, 168
————, conditioned emotional response in, 186–9
————, ear size in, 174
————, effects of supplementation, 20, 154, 184, 211–2, 337, 343, 346–7
————, estimate of chronicity, 282, 317
————, fatfolds in, 338–9
————, fetal, see fetal malnutrition
————, food utilization in, 202–11, 214–5
————, head circumference in, 24
————, head size in, 174
————, height in, 282, 332–4, 342–3, 347
————, hemoglobin in, 235
————, maternal, timing of, 85–8
————, maze running test in, 186–7
————, pituitary in, 188, 191
————, prenatal, see fetal malnutrition
————, rehabilitation, 164, 174–5, 178–9

————, screening for, 16
————, serum iron in, 235
————, skeleton in, 168-71
————, speed of movement in, 153
————, strength in, 154
————, weight in, 282, 332-4, 342-3, 347
Maternal diabetes, 49-55
————, adipose tissue in, 49-50, 52, 55
————, birth size in, 50, 121-2
————, body composition in, 49-55
————, brain maturation in, 55
————, cerebral cell number in, 50
————, cerebral cell size in, 50
————, cerebral water in, 52
————, extracellular volume in, 53
————, fatfolds in, 121-2, 132-3, 280, 338-9
————, adipose tissue in muscle in, 50
————, hematocrit in, 53
————, hormones in, 53-4
————, lean body mass in, 49-50, 53
————, muscle cell number in, 50
————, muscle cell size in, 50
————, muscle water in, 50
————, organ size in, 50, 55
————, placenta, 49
————, water content in cerebrum, 52
————, zinc in cerebrum, 55
Maternal diet, 75, 77, 82, 84, 96-8, 100-5
————, and birthweight, 122, 197-201, 331-2
————, and body composition, 122
————, and brain size, 331
————, and protein supplementation, 193, 195, 202
Maternal malnutrition, 85-8
Maturation, rate of, 252, 328, 346
————, and body composition, 132
————, and body shape, 132
————, and body size, 132
————, and functional capacity, 132
Maturation, sexual, 64, 178
Maze running test, 186-7
Measurement error, 302-5
Menarche, 346
Metabolism, basal, 28

Metabolism, basal, of fat in muscle, 125
Mixed longitudinal studies, 302, 315-6
Mobility, 238
Moiré contourograph, 30-40
Monophotogrammetry, 29-30
Mortality, infant, 243-4, 334, 343
Muscle, and exercise, 48
————, and growth hormone, 54
————, and insulin, 54, 64
————, and somatomedin, 54
————, cell number, 50, 57-8, 60, 64
————, cell size, 50
————, fiber number, 172-3
————, fiber size, 173
————, growth in malnutrition, 48
————, in pubescence, 56-7
————, water in, 50
Multiplex projector, 28

Normative data, see reference standards
Nutritional assessment, 3-4, 8, 310-2
Nutritional needs, 124
Nutritional status, and climate, 232
————, and disease, 4
————, and exercise, 232
————, and psychological factors, 4
————, and season, 4
————, anthropometry, 4, 15-25, 150, 220, 286, 338
————, body composition, 317
————, definition, 109-10
————, estimation of chronicity, 282, 317
————, genetic effects, 239
————, grading systems, 8
————, in individuals, 2
————, in populations, 2-3
————, measures of, 230-1, 238, 244-6

Obesity, 16, 47-8, 55, 236, 255-6, 262-3, 282
————, adipose tissue in, 235
————, aerobic capacity in, 137
————, and breast feeding, 56
————, and diabetic mother, 49-55
————, and exercise, 236, 238

Obesity, and ileal bypass, 60, 64
————, and longevity, 64
————, body composition in, 47, 56-7, 60, 64, 236
————, effects of weight reduction, 60, 238
————, fat cell number in, 55, 62-3
————, fat cell size in, 55
————, fatfolds, see separate listing
————, height in, 56
————, hormones, in 56-7, 60, 64
————, infantile, 2, 10-1
————, insulin, muscle response to, in, 57, 60
————, intracellular mass in, 57
————, lean body mass in, 56-7, 64, 236
————, muscle cell number in, 57-8, 60, 64
————, muscle response to insulin, in, 57, 60
————, sexual maturation in, 64
————, skeletal age in, 58-64
Ovary, 53-60
Overnutrition, see obesity
Overweight, 224

Parent-child correlations, 232-4, 347
Photodermoplanimeter, 29
Photogrammetry, 28-30
————, cyrtographometer, 29-30
————, moiré contourograph, 30-40
————, monophotogrammetry, 29-30
————, multiplex projector, 28
————, photodermoplanimeter, 29
————, stereophotogrammetry, 28
Physical activity, see exercise
Pituitary, 53-4, 188, 191
————, dwarfism, 54
Placenta, in maternal diabetes, 49
————, insufficiency, 75
Plasma amino acids, 3
Prematurity, 77, 79
Prenatal growth, 329
————, and disease, 82
————, and high altitude, 82
————, see fetal malnutrition
Program evaluation, 318
Protein-calorie malnutrition, see malnutrition

Protein intake, and birth weight, 330-2
————, and height, 240, 334, 338, 347
————, and Quetelet index, 240
————, and weight, 334, 338
Protein needs, 1, 4-5
Protein supplementation, 20, 154, 184, 211-2, 337, 343, 346-7
————, and lactation, 195-6
Psychological factors, 4
Pubescence, adipose tissue in, 56-7
————, and exercise, 123, 129
————, muscle in, 56-7
Pyruvic kinase, 95-8, 100-5, 117

QUAC stick, 5-7, 160
Quetelet index (W/H), 240

RNA synthesis, 105-7
Rate of maturation and functional capacity, 130-1
Ratios, see anthropometric measures
Reference standards, 2, 9-11, 223-9, 232, 239, 246-51, 256, 258
————, age classes, 257-8
————, birthweight, 228
————, bivariate, 229
————, height, 229
Rehabilitation, 164, 174-5, 178-9
Relative body weight, 273-81
Residual lung volume, 42
Respiratory distress syndrome, 81
"Runt" pigs, 166-7
————, postnatal growth, 169
————, skeleton, 168-71

Sampling errors, 222-3, 259-60, 292-301
Scales, see apparatus
Seasonal effects, 4, 313, 342
Secular changes, 119-20, 261, 282
Senescence, adipose tissue in, 125
————, and exercise, 123, 139
Serum albumin, 3-4, 8
Serum iron, 235
Sexual maturation, 64, 178

Skeletal age, 58, 64
Skewness, 260
Skinfolds, see fatfolds
Small-for-dates, see fetal malnutrition
Social customs, 48
Socioeconomic status, 4
Somatomedin, 54, 60
Speed of movement, 153
Standards, see reference standards
Stature, see height
Stereophotogrammetry, 28
Strength, 154
Supplementation, see protein
Surface area, 256
Surveillance, types of, 3
Survey purposes, 309-10
Synapses, in fetal malnutrition, 81

Talus inclination, 252-5
Testis, 53,60
Thermography, 93
Threonine, 194-204
Total food intake, 337, 340

Urban-rural differences, 346
————, in body build, 282
————, in body size, 251, 259
————, in difficulty of assessment, 235
————, in foodstuffs, 234
————, in heterozygosity, 238
————, in mobility, 238
Urea nitrogen, 8

Validity of indicators, 316-7
Vital capacity, 252
Vitamin A, 153, 155
Vitamins, blood levels, 3

Water, balance, 17-8
————, total body, 158
Weaning, 15
————, diet during, 337
Weight, 15-8, 21

Weight, and calorie and protein intake, 334, 338
————, and nutritional status, 15-8, 21
————, changes in, 16-8, 22-4
————, for height, see anthropometric measures
————, gain after birth, 55
————, in malnutrition, 282, 332-4, 342-3, 347
————, preschool pattern, 332-4
————, relative body weight, 273-81

Zinc in cerebrum, 55